EMBODYING BIODIVERSITY

biodiversity in small spaces

Virginia D. Nazarea, Series Editor

EMBODYING BIODIVERSITY

Sensory Conservation as Refuge and Sovereignty

EDITED BY

TERESE V. GAGNON

THE UNIVERSITY OF
ARIZONA PRESS
TUCSON

The University of Arizona Press
www.uapress.arizona.edu

We respectfully acknowledge the University of Arizona is on the land and territories of Indigenous peoples. Today, Arizona is home to twenty-two federally recognized tribes, with Tucson being home to the O'odham and the Yaqui. Committed to diversity and inclusion, the University strives to build sustainable relationships with sovereign Native Nations and Indigenous communities through education offerings, partnerships, and community service.

ISBN-13: 978-0-8165-5398-3 (paperback)
ISBN-13: 978-0-8165-5399-0 (ebook)

Cover design by Leigh McDonald
Cover image: Original artwork by Taro Takizawa
Typeset by Sara Thaxton in 10/14 Warnock Pro with Cassino WF and Brandon Grotesque

Library of Congress Cataloging-in-Publication Data
Names: Gagnon, Terese V., 1991– editor.
Title: Embodying biodiversity : sensory conservation as refuge and sovereignty / edited by
    Terese V. Gagnon.
Other titles: Biodiversity in small spaces.
Description: Tucson : University of Arizona Press, 2024. | Series: Biodiversity in small spaces |
    Includes bibliographical references and index.
Identifiers: LCCN 2024005280 (print) | LCCN 2024005281 (ebook) | ISBN 9780816553983
    (paperback) | ISBN 9780816553990 (ebook)
Subjects: LCSH: Human-plant relationships. | Plant diversity conservation. | Agrobiodiversity
    conservation.
Classification: LCC QK46.5.H85 E425 2024 (print) | LCC QK46.5.H85 (ebook) |
    DDC 304.2/7—dc23/eng/20240617
LC record available at https://lccn.loc.gov/2024005280
LC ebook record available at https://lccn.loc.gov/2024005281

Printed in the United States of America
♾ This paper meets the requirements of ANSI/NISO Z39.48-1992 (Permanence of Paper).

To Virginia Nazarea, for teaching us that scholarship can be caretaking relations. Thank you for reminding us that our work is the beautiful and serious business of co-completion.

INVOCATION

Let the moon wobble.
Let the basil plant flower.
Let the poets discombobulate.
Let the verbs noun.
Let the nouns verb.
Let the grief howl.
Let the emails unread.
Let the land speak.
Let the oceans revenge.
Let the people free.
Let the people free.

—Ally Ang

CONTENTS

Part II. Sovereignty

FOREWORD I

While just three global nonprofits get funded to do Conservation with a capital *C* to the tune of US$1 billion per year, most of their donors do not understand that many kinds of conservation with a small *c* have effectively conserved and restored wild plants and their habitats and cultivated plants and the cultural landscapes for millennia. As farmers, herders, gardeners, and foragers practice carbon pulldown in their silvo-agroforestry systems with a lowercase *c*, they get virtually no intellectual credit nor carbon market credits either. But Terese Gagnon's *Embodying Biodiversity* aspires to set the record straight, and perhaps even move the needle in the other direction.

That other direction has ethnic ethics embodied in every meal, engendering a kind of conservation you can taste and savor. It is a more participatory, collaborative, place-based, faith-based, and biocultural conservation model that will continue after the mega-grants from Carlos Slim, Jeff Bezos, and Bill Gates are closed down and the principal investigators move on to another fashionable topic. It is a kind of ethnoecological restoration that restories a community with more parables even as it restores habitats with little-known food plants, thick-billed parrots, or lesser long-nosed bats. It is humble but lasting; vernacular rather than academic; personalized rather than generalized or essentialized.

The metaphorical gifts that Terese Gagnon and her mentor Virginia Nazarea have offered us refresh, revive, and celebrate ancient practices of stewarding plant–people relationships for the long haul. We can evaluate a grassroots initiative by how it makes a habitat smell, feel, and function, as well as by how the food on the table expresses unparalleled flavor, texture, and fragrance.

◇◇◇◇◇◇◇◇◇◇◇◇◇◇

Most plant varieties domesticated on this planet were carefully and caringly selected not by PhD geneticists but by peasant farmers and gardeners, and that trend will likely continue in the future, as it is now with cannabis, aya-huasca, Aunt Molly's ground cherry, Chimayó chile (or chile de Cobán), and Brandywine tomatoes—as well as maize, bananas, ensete, and more. The way Gagnon elucidates the practices of a half dozen farmers and gardeners should be required introductory reading for anyone who dreams of becoming a botanist, plant breeder, or conservation biologist.

To top it off, Gagnon gets a prize for the most evocative and refreshing early-career prose I have ever read. Let it slay like a dragon all the stuffy, jargon-laden, self-referential academic fluff coming out of universities at a rate that depletes one forest after another year after year.

I vote with Terese for conservation you can taste with a capital *C*.

—Gary Nabhan

FOREWORD II

At a time when the growing dominance of monocultures increasingly threat-
ens our future, this collection poses a direct challenge to the violent ex-
tractions of settler colonialism and capitalism, and to the complementary
abstractions of modern epistemologies. The distancing abstractions are
directly opposed to anthropological ways of knowing, which offer context-
dependent understandings in response to the context-free knowledge claims
of modernity and embedded holistic meanings as opposed to the disembedding of economy from society that characterizes neoliberal capitalism.
While capitalism continues to objectify nature and alienate human labor,
this collection counters this and further reflects growing concern with the
senses and the body as sources of knowledge of the interrelation between human and more-than-human worlds. As Gagnon writes in the introduction:
"People's choices to maintain good relations with certain plants exceed the
logic of rational use value. Rather, such choices are often rooted in sensory
experience and affect."

This collection is dedicated to Virginia Nazarea and shows the multiple
strands of her influence. I can think of no other scholar who has done more
to bridge studies of agrobiodiversity with the anthropology of food, production (growing), and consumption (cooking, eating) than Nazarea. From
her initial insights, building from her work in cognitive anthropology that
ethnoagronomy requires ethnogastronomy, Nazarea focused our attention
in laser-like fashion on how the choices of farmers are based not only on
questions of markets and efficiencies, but equally (sometimes much more)
on the sensory dimensions of crops, their tastes, and their appropriateness
for different occasions (a snack, a child's first mouthful, a food to eat when

rice is present or absent); all these diverse and contingent criteria are without measure, nor plottable on a graph, and they make biodiversity continue in our world. Nazarea was among the first to incorporate *memory* into ethno-agronomy, showing how collective cultural memory is necessary for but-tressing biodiversity against all the forces that make it disappear. But, just as important, she showed that biodiversity is necessary for people to live meaningful lives, that life is incomplete without the variety that enables an embodied sensory appreciation of the different important social occasions that resist modernity, as in Anderson's (1983) "homogenous, empty time" or the "forced amnesia" of various totalitarianisms (Connerton 2009).[1] Nazarea asks us, even when not faced with power and oppression, to pay attention to the pockets of memory in the minds of farmers who "poke fun at facile generalizations and challenge revered theodicies about agriculture, about development, and about the whole strategy of 'empowerment' and integra-tion" (1998, 115), or in the minds of Peruvian potato farmers, who treat their crop as "'infants' that they swaddle in blankets and sing to, wily ones that jump out of the ground to greet them, and errant ones they scold before taking back in" (2021, 264).

Nazarea emphasizes the importance of diverse types of conservation, not just the "cold" variety, which remains necessary to preserve genetic mate-rial under threat from wars and environmental disasters, but also everyday modernist practices, which are often treated as the only legitimate, scientific conservation practice. Nor does she praise only in situ conservation, which is "warmer" but can still fall into the abstracting tendencies of any program-matic scheme. Rather, she asserts that this diversity must also include the "heat" of in vivo conservation and the local knowledge it (re)produces, as well as the equally important trans-situ conservation, represented by the movement of people and their seeds across borders and boundaries cre-ated by those same world disasters, as people caught up in the midst of change pursue their "desire to preserve and reconnect with flavor memories from home," which in turn "compels seeds' international travel and preser-vation" (Ramsey). These conditions are what lead people to seek sanctuar-ies, or meaningful coherences that enable them to continue amid plantation landscapes (Novak), or to find refugia, as in the dreams of Ugandan banana cultivators who, in the words of one such farmer, "used to plant trees to remember our events [since] . . . in the past, we did not have anything to write on" (Sato). These refugia also take the form of metaphorical seeds of future sovereignty carried by Karen refugees as sensuous stories (Gagnon), as important as the literal trans-situ conservation they practice in creating the memories and choices that, in Nazarea's words, have not been erased by "forced amnesia"; as important as the packages of seeds carried along with the food from home by El Salvadoran migrants in Colorado, which help, along with their memories of pre–green revolution practices, to reconstruct a sustainable agriculture amid various forms of violence (Anastario).

The notion of sensuous sovereignties is an important addition to the conceptual apparatus Nazarea has brought together, suggestive as it is of embodied knowledges, which create "open-air schools" in homegardens for the exchange of memories and situated learning (Ibarra, Caviedes, and Barreau); differentiate between the deep cultural meanings of West African trans-situ conservation of Carolina gold rice and the "top-down, production-driven, disembedded approach of genetically modified agriculture of 'Golden rice'" (Simpson), just as with the contrast between the inscribing practices and "disembodied level of the global (the view from nowhere)" of Ethiopian fast-growing trees and the furtive conservation of incorporated practices of ensete home gardens (Peveri).

In Ramsey's trenchant summary of her research among Latinx farmers in Florida, she notes that "engaging with landscapes, plants, and the foods they produce draws on all the senses, calling forth an almost artistic means of coming to dwell in new landscapes and homes. Sensory and embodied engagements help root immigrants in new locales, even as they preserve important ties to people, landscapes, and memories from home." Here I am reminded of how the past thirty years of anthropological research on transnational migration have provided a whole new paradigm for exploring the movement of people, ideas, and nonhuman objects, enabling us to rethink the ways in which spaces become landscapes of meaning. With this collection's drawing on and continuation of Virginia Nazarea's groundbreaking ideas, we are indeed reminded of not only how "reciprocity between plants and humans has existed for millennia and has been essential to human life" but how much can be learned by tracing this reciprocity from the growth of seeds to the dreams and memories of commensality.

—David E. Sutton

Note

1. Taken from my interview with Virginia Nazarea in 2022. See Sutton, "'This Sweet Potato is Beautiful': From Ethnoagronomy to Ethnogastronomy in the Work of Virginia Nazarea," *Food Anthropology*, March 28, 2022, https://foodanthro.com/2022 /03/28/this-sweet-potato-is-beautiful-from-ethnoagronomy-to-ethnogastronomy -in-the-work-of-virginia-nazarea/.

References

Anderson, Benedict. 1983. *Imagined Communities: Reflections on the Origins and Spread of Nationalism*. London: Verso.

Connerton, Paul. 2009. *How Modernity Forgets*. Cambridge: Cambridge University Press.

Nazarea, Virginia D. 1998. *Cultural Memory and Biodiversity*. Tucson: University of Arizona Press.

Nazarea, Virginia D. 2021. "Ontologies of Return: Terms of Endearment and Entanglements." In *Moveable Gardens: Itineraries and Sanctuaries of Memory*, edited by V. Nazarea and T. Gagnon, 253–72. Tucson: University of Arizona Press.

EMBODYING BIODIVERSITY

Introduction

Wonder at the Ends of Worlds

TERESE V. GAGNON

> You do not have to be good.
> You do not have to walk on your knees
> For a hundred miles through the desert, repenting. You only have to let
> the soft animal of your body
> love what it loves.
>
> —MARY OLIVER, "WILD GEESE"

This volume is about the deep interconnections between plants, in all their astounding diversity, and humans. The chapters included here explore the living spheres of everyday plant biodiversity conservation otherwise known as conservation with a small *c*. These realms of conservation are rooted in sensory experiences and embodied connections that sustain reciprocal relationships between plants and humans, as anthropologist Virginia Nazarea has described (2005).[1] This contrasts with the prevailing global narrative, focused on formalized, top-down conservation, or Conservation with a capital *C*.

Large-scale initiatives such as the creation of parks and reserves and the signing of international agreements are often seen as the hallmarks of biodiversity conservation. The authors of this volume contest this narrative by attending to the power of sensory-driven, or sensuous, conservation. We also assert that it is impossible to separate the persistence of diverse plant lives from our human lives, as they are differently embodied and made meaningful in entangled and rapidly changing worlds. Following scholars such as Arturo Escobar (2018) and Marisol de la Cadena and Mario Blaser (2018), who write about the concept of the pluriverse, or how different knowledges and practices *make worlds*, we speak here of *worlds* in the plural.

Top-down Conservation has largely been governed by the logics of Enlightenment thought, which makes what John Law (2015, 128) calls the "one-world world." In reality, far from being an omniscient gaze from nowhere (Haraway 1988, 581) as it imagines itself, the one-world world and its concepts of nature are rooted in western European traditions of intertwined capitalism and Christianity (Cronon 1996, 10–15). In contrast, everyday sensuous conservation is motivated by a vast plurality of ways of knowing and feeling. The everyday and embodied realms, far from demanding ontological purity, allow ample room for heterogeneous means of relating and coflourishing with other species.

In this volume, we explore how everyday sensuous conservation not only complements top-down efforts at Conservation but in fact makes up the majority of conservation worldwide (see Nazarea 2005). These chapters illustrate how sensuous conservation enlivens relationships between plants and humans rooted in mutual dependency, affect, and love. These reciprocal relationships, or kinships, are steeped in place. They are carried out via the mycelial networks of daily life. They are agential and intentional as much on the part of plants, and plant mindedness, as on the part of humans (Ryan, Vieira, and Gagliano 2021).

Reciprocity between plants and humans has existed for millennia and has been essential to human life. Yet, in recent centuries, generalized reciprocity between plants and humans has been overshadowed by the voracity of racial capitalism's extractive relationships. Such bad relations have depleted wild plant populations, homogenized agrobiodiversity, and exhausted soils to the point of losing fecundity. As it turns out, negative reciprocity is a recipe for starvation, as Robin Wall Kimmerer soberingly describes: "In the old times, individuals who endangered the community by taking too much for themselves were first counseled, then ostracized, and if the greed continued, they were eventually banished. . . . It is a terrible punishment to be banished from the web of reciprocity with no one to share with you and no one for you to care for" (2013, 307).

According to the Food and Agriculture Organization of the United Nations, 75 percent of agrobiodiversity worldwide was lost in the twentieth century alone (FAO 1999). Partially as a result of monocultures and the pollution caused by globalized industrial agriculture, many ecosystems are on the verge of collapse. As Leonardo Figueroa-Helland and co-authors state, "the globalized modern-industrial food system not only fails to feed large segments of the population; it also destroys ecosystems, soil fertility, and biodiversity, drives climate change, displaces communities, and erodes cultural diversity. By 2012 genetically homogeneous 'modern' monocultures already covered roughly 80 percent of global arable land" (Figueroa-Helland, Thomas, and Pérez Aguilera 2018, 176–77).

In these times of ecological grief, as one after another of our species cohort mates depart, there is a critical need for a groundswell of what Kim TallBear calls more-than-human caretaking relations (TallBear 2019, 25). TallBear contrasts caretaking relations with American dreaming, which is built on narratives of inclusion within liberal nation-state futures. American dreaming includes some while violently excluding those deemed others. Unlike worlds of mutual flourishing knitted by caretaking relations, nation-state dreaming creates worlds of division and death.

Historically, a relative few have profited from the narratives of nation-state dreaming. Relatedly, the responsibility for driving climate change and biodiversity loss lies principally with a specific subset of humanity: a far cry from the singularly culpable *anthropos* of the Anthropocene (see Bauer and Bhan 2018, introduction). As Janae Davis and co-authors write, "The Anthropocene is clearly not the product of 'human nature,' or humanity as a whole, but rather interrelated historical processes set in motion by a small minority. This privileged cadre provided the preconditions for the development of global capitalism through processes of settler colonialism and enslavement, organized and rationalized by racism" (Davis et al. 2019, 4). As such, scholarship that seeks to attend to more-than-human relationships must *contend* with the vast numbers of people who struggle every day to retain their tenuous spot in the realm of humanity.

In the face of global injustice, inroads toward caretaking relations may include modes of feeling with the pulsating liveliness of the world and other beings. Rather than being a diversion from a linear path toward justice, reconfiguring our senses can help us feel into right relations. In this vein, the chapters in this volume highlight instances of caretaking relations emanating from numerous parts of the world and emerging across diasporic routes. These chapters offer glimpses of possibilities for nurturing good relations between plants, humans, and other beings amid a rapidly changing biosphere—also known as the places we call home and our own bodies.

The scholarship included here demonstrates that everyday sensuous conservation is often found in unassuming places. Such quiet spaces of biodiversity conservation range from small-scale farms and homegardens to kitchens and temples. In situations where freedom and rights are profoundly constrained—such as in securitized refugee camps and via treacherous migration routes—there is evidence that people still find creative ways of maintaining relationships with their plant kin. This includes traveling with foods and seeds, or simply keeping alive stories from home related to plants. Such liminal spaces are often critical boundaries in struggles between biocultural continuance and loss.

The chapters in this book show that everyday sensuous conservation does not only preserve germplasm. Spaces of living conservation also transmit knowledges and embodied practices that facilitate the cultivation of plants and their use for culinary, medicinal, artisanal, and spiritual purposes. As Virginia Nazarea discusses, the "warm milieus" where taste and stories are centered are critical to the living conservation of biodiversity (see Nazarea 2021, 266 and 268). People's choices to maintain good relations with certain plants exceed the logic of rational use value. Rather, such choices are often rooted in sensory experience and affect. Attention to the lives of plants and reciprocity with them generates love. Love and its sibling, care, provide possible modes of moving through the unfolding ecological tragedy of deeply differential human making.

Caught in the Fabric of the World

> Visible and mobile, my body is a thing among things; it is one of them. It is caught in the fabric of the world, and its cohesion is that of a thing.
> —MAURICE MERLEAU-PONTY, "EYE AND MIND"

What does kin making mean in troubled political and ecological times? One thing it means is vulnerability: knowing that those we love and depend on may be lost, and choosing to love and depend on them anyway. Environmental political theorist Stephanie Erev (2019) describes how our human bodies are bound up with what is already being transformed by climate change. She illustrates how we sense these shifts on a subliminal level even while we lack the grammar to make sense of such "bodily feelings" (2019, 837). How then, she asks, might we lean into our animal intuition, which perceives these changes, to find different ways of relating to the world?

Following the poet Mary Oliver, we can choose to embrace the soft animal inside us. Through practices of attention and wonder, we can learn to feel ourselves as connected to the materiality around us. When we do this, we can comprehend climate change and biodiversity loss as something happening to our own bodies. Indeed, we are always already more-than-human. Carol Wayne White (2017) demonstrates this through an excavation of the planetary origins of our political agency. Digging into deep time and our cellular beginnings, she shows how humans are not just drivers of destruction and harm, external to natural systems. Rather, we are of these systems. In place of human exceptionalism, understanding our being as fundamentally connected to the rest of the world may help us reconfigure our relationship with it.

These times of ecological grief necessitate a multiplicity of knowledges, as Robin Wall Kimmerer attests (2013). This volume highlights such plural and overlapping ways of knowing and relating to plants. Across these chapters, we witness how Enlightenment ways of knowing plants are frequently insufficient on their own. Instead, people in various places and circumstances find value in complementary modes of sensing that help them order their responsibilities to other beings. These beings include plants, animals, fungi, soils, waters, stones, human relatives, deities, ghosts, ancestors, and more. Here we sense that our knowing is bodily. We are made of the same stuff. Just as the cloud is in the cup of tea, so too the rice plant, the corn silk is in our hands.

Origins and Contributions of this Volume

This volume came about through sustained conversations around themes of biodiversity, embodiment, and sensory scholarship between the authors and others. This book is greatly influenced by the scholarship and mentorship of Professor Virginia Nazarea. She has been the spider in the relational web connecting us. More specifically, this volume was born out of the invigorating dialogues begun at two conference panels co-organized by Virginia Nazarea and Terese Gagnon. The first was a panel in November 2018, "Refuges of the Blighted Wilds," at the Symposium of Australian Gastronomy, held on the unceded lands of the Burramattagal people, otherwise known as Paramatta, Australia.[2] The second was a panel in November 2019, "Sovereigning the Sensuous: Sanctuaries in Charged Climates," at the American Anthropological Association annual meeting, which took place on the unceded lands of the xʷməθkʷəy̓əm (Musqueam), Sḵwx̱wú7mesh (Squamish), and səlilwətaɬ (Tsleil-Waututh) Nations, otherwise known as Vancouver, Canada.

Our aim in crafting this book is to highlight the sensory nature of biodiversity conservation, rooted in local places and across diasporic routes. We are inspired by the tenacious, creative work of farmers, seed savers, and activists who commit their labors to countering agrobiodiversity loss and violent food systems while nurturing flavor, seed stories, and communities. These practitioners of collective nourishment embody the tremendous possibilities that exist for a resurgence of restorative food systems and tasty foods.

This interdisciplinary work argues for the importance of everyday sensuous conservation and its ability to grow diverse, livable worlds where human embodiment is understood as part of—not separate from—plant life. We begin from the understanding that the vast majority of biodiversity conservation worldwide is carried out not by large-scale, top-down conservation projects but rather by ordinary people who engage in sensory, place-based relationships with specific plant species. Collectively—drawing on diverse methodologies and research localities spanning multiple continents—the authors highlight the vital role of embodied experience in forging enduring relationships with plants. The book argues that the genius of ordinary people in stewarding biodiversity should be recognized as such. This recognition is crucial to combating ecological harm and global injustice stemming from racial colonial capitalism, which enacts "the one-world world" (Law 2015).

The wide array of disciplines this volume speaks to is one of its major strengths. The diverse methodologies and geographies spanned by the chapters illustrate the vivid array of sensuous conservations and approaches to studying them. Additionally, a key scholarly contribution of the volume is bringing ethnobiology and multispecies ethnography approaches into dialogue. Virginia Nazarea's scholarship, which has inspired the authors, represents the groundbreaking convergence of these two related areas of study, which too often speak past each other. By bringing multispecies ethnography and ethnobiology scholarships into direct conversation, this book reveals that plant biodiversity is not a resource or a service but a part of people, one that gives their lives meaning and wholeness.

Embodying Biodiversity builds on and furthers the aims of the preceding volume co-edited by Terese Gagnon, with Virginia Nazarea, *Movable Gardens: Itineraries and Sanctuaries of Memory* (2021), while making its own contribution. The present volume continues the project of attending to storied and affective relationships between people and plants in an era of planetary unmaking. While *Movable Gardens* is principally about modern conditions of displacement and peoples' attempts at emplacing with plants, the present volume focuses on the interdependent physicality, the very embodiment, of people and plants. It traces how these are interlaced with sensory, storied worldings that counter the worlds-destroying force of racial colonial capitalism.

The collection is divided into two connected parts. Part 1, Refugia, explores how, amid charged climates and systems of oppression, embodied connections with plants are maintained in spaces of sanctuary from which they may later be *reseeded.* Part 2, Sovereignty, focuses on how the intangible aspects of humans' relationships with plants become inscribed and made legible through processes of *sovereigning the sensuous.*

Part 1: Refugia

Just as our troubled ecological present calls for diverse modes of relating, so too does it necessitate diverse seed stocks to meet the significant challenges of climate change adaptation. This may include plant genetic materials considered new, or "foreign," to certain places. Plants and people alike are being forced to shift their homes at historically unprecedented rates, in some cases moving together.

In fact, all migration is comigration when it comes right down to it. As humans, about half the cells in our holobiont bodies are bacterial (Gilbert 2017, M75).[3] We shape our home environments through our microbiological aura, which travels with us from place to place (Lax et al. 2014). Additionally, there is the quietly ubiquitous phenomenon of people making home on the move by traveling with plants, seeds, and foods (see Keeve 2020; Nazarea and Gagnon 2021). These practices have persisted across time and space, even in the face of great adversity and strict border regimes.

The stories included here provide glimpses of what more-than-human collaborations for survival look like amid ecological destruction and in the teeth of violent oppression. Some of the chapters illustrate how caretaking relationships are tended on the edges of racial capitalism's seemingly endless expanse. Davis and co-authors assert that "plotting within and against the plantation is a practice of cultivating life and kin that challenges the intertwined death-dealing logics of racism and ecocide" (Davis et al. 2019, 3). In this vein, we witness plotting as the embodiment of nourishment in the long wake of the plantation (see also Thomas 2019). Relatedly, chapters in this volume offer insights into escape agricultures in which people and plants collaborate for escape—from state control, war, and dispossession. In such contexts, as James Scott describes, "in place of concentrated grain crops that ripen simultaneously, you would prefer shifting, diverse, dispersed, root crops of uneven maturation" (2009, 178). These strategies provide refugia (Tsing 2017, 54), or spaces of sanctuary, from which future generations of intertwined life might be *reseeded*.[4]

Refugia emerge in the choices of Salvadoran farmers to maintain centuries-old maize-cultivation practices in the face of ongoing epistemic violence (Anastario). They become perceptible through transnational ritual practitioners keeping alive plant kinships in devotion to the goddess Mariamma within postindustrial urban landscapes (Novak). Creative forms of refugia are present when Karen refugee youth tell sensory-laden stories about the plants they foraged near their home villages, which they crave while living in a refugee camp (Gagnon). Refugia are present when banana cultivators in Uganda dream about the banana plants situated on specific parts of their property, not in the abstract but in the particular (Sato).

Mike Anastario analyzes epistemological entanglements between traditional and scientific ways of knowing maize in the highlands of El Salvador. He describes the roles that observing the behavior of birds, dreams, and lunar cycles play for farmers in predicting when the rains will come, or when they should plant and harvest their crops. He digs into the histories of violence, including that of epistemicide, by which traditional modes of knowing have been suppressed. This violent suppression continues into the present, for example, taking the form of agricultural extension agents who promote scientific ways of knowing while actively discouraging traditional practices of maize cultivation.

In the wake of the Salvadoran genocide and amid ongoing epistemicide, the farmers Anastario works alongside continue relying on and passing down traditional methods of maize cultivation. For the farmers, these practices sit alongside the techniques taught by modern agricultural science. In fact, farmers who continue to rely on these entangled ways of knowing maize hide traditional practices among those of modern agricultural science and thus protect them.

Shannon A. Novak, in her chapter, traces the movement of devotees to the goddess Mariamma with the plants essential to their ritual practice. This history of comigration and plotting against the plantation takes us from

South India to British Guiana, and finally to a suburb of Toronto, Ontario, known colloquially as Flower City, where Novak's interlocutors reside and worship. She explores how the embodiment of ritual practice at the temple, including its intense smells of smoke and flowers, the sounds of drums, and the rhythm of dancing, intersect with the transnational movement of plants and thus the preservation of biodiversity through association with the divine.

Novak digs deeply into the history of the community and their movements, along with their plants, to examine tangled relationships between plantation and plot across time and place. Embodied engagements with plants are particularly important here, as many of the ritual practices are dependent on specific plants. Some can be grown locally in the city's northerly climate, but many, such as neem, must be imported from the Caribbean or other tropical places at considerable trouble and expense. On the fringes of a postindustrial city, these ritual practices provide a refuge from the practitioners' daily working lives that extends the legacy of the plot as a site of resistance and nourishment in the long wake of the plantation.

Terese V. Gagnon explores how Karen young people living in Mae La refugee camp on the Thailand-Myanmar border refuse dispossession from their homes and the more-than-human relationships tended there. They enact their refusal by imbibing foods from home sent across the border. They additionally resist dispossession by telling sensory-rich stories about agriculture and foraging in their home villages.

Telling stories and tasting foods that recall more-than-human relationships enact a movement toward the restoration of these desired conditions. These young people demonstrate that nothing that can be recounted is truly lost. In the refugee camp, where residents are compelled to become "self-reliant" by aid organizations, the actions undertaken by these refugee young people challenge humanitarian notions of life as survival. Acts of sharing food and of dreaming aloud about future returns to their "native places" refuse the erasure of Karen sovereignty and preserve the seeds of more-than-human caretaking relations, even in exile.

Yasuaki Sato writes about the relationships between bananas and their people. More specifically, he explores Ugandan banana farmers' embodied connections with the diverse banana plants grown in their homegardens. He describes how farmers contend with the presence of diversity among their banana plants. Of particular interest is Sato's investigation of how people treat "new," or unnamed, varieties of bananas, which he found to have almost a supernatural quality: being either a blessing or a curse. Respectively, these "new" plants were either investigated to death by curious neighbors or abandoned and left to die as something unnatural, an aberration.

Additionally, Sato describes how his interlocutors have fortuitous dreams predicting what will happen with specific plants on their land. He also details the banana identification competitions he organized as part of his research method, in which he discovered how knowledge of different banana varieties manifested differently across categories of age and gender, with the accrual of time and relationship to the land being important factors. Sato found that women in Uganda have a special relationship to banana plants. Most significantly, he relates how the Ugandan banana farmers he collaborates with rely equally on their observations as they do on established facts about how certain banana varieties look, taste, produce fruit, and otherwise behave. In this way, people can hold seemingly contradictory ideas about bananas without epistemic conflict. As such, in the realm of bananas, two things can be true at once.

Part 2: Sovereignty

Anthropologist Heather Paxson, writing about more-than-human relationships in artisanal cheesemaking in the United States, observes how when things scale up, they change *qualitatively* (2012, 138–51). This means trading connections with the sensory and embodied dimensions of labor as business models and profit margins get bigger. Does the same hold true for relational connections between persons and landscapes as political systems scale up?

Sovereignty and the sensuous are, on the face of things, opposites. One involves embodied connections and situatedness, while the other involves formalizing processes that turn landscapes into territory. Yet the two are connected. Embodiment arises from dwelling in place and sensory engagements with one's surroundings that accrue in the memory. As such, what happens when the embodied realm seeks legitimacy, permission to move from the sensuous to the sovereign?

Pertinent to connections between sensory experience and sovereignty is the fact that Indigenous communities steward 80 percent of the Earth's biodiversity while making up only 5 percent of the world population (FAO 2017). Yet, Indigenous peoples, having survived centuries of genocidal violence and ongoing attempts at erasure from settler colonial states, are notably excluded from decision-making processes about biodiversity conservation on the world stage. What insights might these facts provide into the nature of sovereignty and the motivating forces for conservation? Some of the chapters in this volume address how Indigenous sovereignty persists even amid violent oppression, though not without considerable loss.

Furthermore, when we speak of sovereignty as related to embodiment and economies, we might heed J. K. Gibson-Graham's (2006) call to cultivate means of noticing and describing not only capitalist relations but also counter-capitalist and noncapitalist forms of relating. Such self-sovereign spheres are already being tended on the fertile margins of dominant political, social, and economic systems. Several chapters in this volume attend to such spaces.

Sovereignty emerges in the form of thriving homegardens in the southern Andes, and the diverse communities they assemble, which are as much the treasuries of agrobiodiversity as gene banks. Yet these homegardens are additionally equipped with diverse and deep ways of knowing the earth. Their presence within local landscapes facilitates the sharing of both germplasm and knowledges (Ibarra, Caviedes, and Barreau). Sovereigning of the sensuous can be observed through storied accounts of Latinx farmers seeking legibility on their own terms within data-driven agroeconomic landscapes of the United States (Ramsey). Countervailing registers of sovereignty are present in Ethiopia, when political rhetoric around the planting of fast-growing trees is contrasted with the largely hidden, almost furtive nurturance of biodiversity in backyard gardens (Peveri). Sovereignty shimmers through rice worlds that assemble alternative textures of the future: from those rooted in modernist technofixes to others simmered in the complex waters of taste and place (Simpson).

José Tomás Ibarra, **Julián Caviedes**, and **Antonia Barreau** investigate the agrobiodiversity of homegardens in an intercultural context in Wallmapu, the ancestral land of the Mapuche people, also known as the southern Andes. Their findings point to increased overall diversity of plant species and varieties present in homegardens as a result of interaction and exchange between Mapuche and non-Mapuche campesinos, as well as lifestyle migrants to the region.

The authors' participatory research melds theory and practice, by facilitating workshops where gardeners of different backgrounds exchange seeds and agroecological knowledges. In this context, embodied connections with diverse plant species grow out of local integration and overlapping ontologies. In particular, the authors uncover the power of seed, plant, and knowledge exchange among campesinos and lifestyle migrants to enhance the flourishing of agrobiodiversity and resiliency, with possible wider implications for territories experiencing demographic and socioenvironmental changes.

Emily Ramsey explores the relationship between market access and embodied practice for Latinx immigrant farmers in the United States. She highlights the trend of *datafication* in farming—in which all aspects of farming and farm business become subject to flattening and legibility as data. She focuses on U.S. agricultural extension agents' emphasis on datafication when leading training sessions aimed at Latinx farmers. At such sessions, farmers are encouraged to make themselves *legible* to the state and the Market. Ramsey guides our attention as readers away from the capital *M* Market, and to numerous lowercase *m* markets. She highlights these lowercase *m* markets as localized spaces where commercial viability and embodied practices can meet in relative harmony.

Through insightful ethnography, Ramsey details how these Latinx farmers, who have begun their own farm businesses, navigate the space between access to markets that allow them to earn a living and the embodied joys of farming as an intrinsic value, which motivated them to farm in the United States in the first place. In contrast to the data-driven vision of farming that constantly asks farmers to scale up and to get big fast, Ramsey's interlocutors assert their own vision of what successful commercial farming looks like.

Valentina Peveri explores the chasm of difference between large-scale, top-down Conservation projects in broad public view and embodied conservation in private, unassuming spaces. In the context of Southwest Ethiopia, Peveri delves into the performative and politically motivated planting of fast-growing trees, heavily documented in social media posts. This is contrasted with the almost furtive conservation that happens in backyard ensete homegardens.[5] While the tree-planting initiative centers monoculture and high-speed results, these hasty trees end up disrupting existing ecosystems. Conversely, in the edible homegardens of Peveri's interlocutors, kin relations between plants and people are nurtured without showiness.

In a world seemingly fixated on productivism and instrumentalist approaches, these backyard homegardens are rooted in affective connections. Amid the hypervisual, performative environmentalism of social media and

PR campaigns, ensete homegardens are intentionally made *illegible*. They are motivated by intrinsic values, including taste, embodied joy, and connection. Paradoxically, as Peveri demonstrates, for all the pomp of tree-planting projects, the agrobiodiversity benefits arising from the warm, earthy spheres of backyard gardens are profoundly more meaningful. Peveri's ethnography is interwoven with her experiences of sensory deprivation and mediated life during the height of the COVID-19 pandemic. These reflections achingly drive home the message that the embodied realm is not frivolous but essential to more-than-human sociality.

Justin Simpson writes of the contrasting ontologies of two rice varieties, golden rice and Carolina gold rice. He uses these rices and the worlds they assemble to present a new materialist theorization of biocultural diversity. On the one hand, golden rice, a magic bullet crop aimed at eliminating vitamin A deficiency through genetic engineering, assembles a world of high modernism. Yet its technofixes are highly targeted and fail to consider the underlying causes of the malnutrition golden rice is intended to address. Instead, golden rice leaves new biocultural challenges in its wake. On the other hand, Simpson presents the heritage breed, Carolina gold rice. This rice is a legacy of the rice cultures and skills of West Africans, who were stolen from their homes and brought to the southern United States under chattel slavery. After almost disappearing from use following a hurricane, Carolina gold has experienced a renaissance in recent decades, becoming favored by hunters, chefs, and environmentalists alike.

Foregrounding the violent history of slavery and white supremacy of Carolina gold rice, Simpson stays with the trouble of this noninnocent varietal. Simpson describes how Black chefs in the U.S. South, such as Chef Dennis, work with Carolina gold rice as a way of reviving and sharing Gullah Geechee foodways, which anchor people to their past and bring back a kind of self-sufficiency. Here, seemingly small acts, such as networking and seed sharing, are also a revolution. Simpson shows how rice is not just a crop but is connected to entire ways of being in the world: that rice cultivation is ontological labor.

Conclusion

Collectively, this volume attests to the importance of reciprocal relations between plants and humans, as they emerge in small everyday spaces. These relationships are undergirded by sensory and embodied connections, which overflow the calculus of rational use value. In many cases, these relationships extend beyond physical realms, stretching into the realms of dreams, devotional practices, and visions of pasts and futures. This collection also evidences how these relationships can be sustained along complex transnational routes and in the face of hegemonic forces that attempt to minimize or trivialize their existence.

As climates shift, set notions of which plants and which people belong where are inevitably changing. Humans, along with plants and trees, are increasingly becoming refugees, forced to flee to habitable zones. The prevailing response to such movements globally has been the securitization and militarization of borders and techniques of containment that seek to keep human and more-than-human others out. Such expulsions of human relatives from the realm of humanity are fundamentally intertwined with the logics of colonialism and racial capitalism, which are the drivers of climate crisis and the collapse of ecosystems.

Instead of hurtling progress toward death and ecocide, what might caretaking relations look like under such conditions of change? The chapters of this volume point toward spaces of everyday embodied conservation, which leave room for heterogeneity, as some possible answers to this vital question. The scholarship included here ultimately implicates us, asking us to consider how we understand ourselves as connected to the beings we travel through life with, and on whom we depend. It implores us: How might we contribute to the healing of wounded places in the world?

In sitting with these questions, rather than allowing them to remain rhetorical, I attempt answers for myself. Moving in the spirit of this volume— from the located and particular rather than the universal—I humbly share my experimental answers here and invite you to do the same. To really answer these questions. To record your thoughts. To return to them again and again.

Consider these speculative responses to the questions: What might my caretaking relations look like under conditions of global change and unraveling? How do I understand myself as connected to the beings I travel through life with and depend on? How might I (right now) contribute to the healing of wounded places in the world?

I understand that community care is as important as new knowledge production. I honor the fact that it is okay, even vital, to acknowledge that we are racinated beings, beings with roots, with connections and obligations to specific humans and more-than-humans. "Being productive" should not come before my participation in dense networks of care or before giving and receiving support; before experiencing joy and pleasure. These are essential.

I recognize and celebrate the fact that the world is not principally a text. There are other modes of sharing understanding, of inscribing, of dancing with the world. Knowledge production and even meaning making are not the only valid forms of creation. Other generative forms of creating can be emotional, affective, and environmental. They can be bodily and sensory. They can be acts of composting, sowing, growing, sewing, cooking, feeding, mending, healing, making music, praying, dancing, eating, laughing—and otherwise nourishing and being nourished.

Emplacing and loving are labors I attend to. My and our survival depends on it.

Notes

1. Our collective attention to biodiversity in small spaces and embodied experience has been strongly influenced by the work of Professor Virginia Nazarea. This volume is inspired largely by her scholarship and generous mentorship. It is a celebration of the attunements she has imparted to us and many others.

2. The place name meaning "place where the eels lie down."

3. Scott F. Gilbert defines a holobiont as "an organism plus its persistent communities of symbionts" and provides the example of a cow plus the symbiotic community of microorganisms in her gut "that digests the grass and makes the cow possible" (2017, M73).

4. The concept of *refugia* comes from population biology. Anna Tsing further theorizes this in relation to present ecological destruction. Tsing suggests that refugia are not only protected pockets where genetic diversity can persist amid threat but includes biocultural practices within her configuration. Such refugia have the potential to repopulate wider publics after periods of profound loss. But Tsing warns that we are approaching a point in advancing ecological destruction where even refugia may be unable to be sustained.

5. As Peveri describes in detail, ensete is a root tuber crop native to Ethiopia.

References

Bauer, Andrew M., and Mona Bhan. 2018. *Climate Without Nature: A Critical Anthropology of the Anthropocene.* Cambridge: Cambridge University Press.

Cronon, William. 1996. "The Trouble with Wilderness: Or, Getting Back to the Wrong Nature." *Environmental History* 1 (1): 7–28. https://www.journals.uchicago.edu/doi/10.2307/3985059.

Davis, Janae, Alex A. Moulton, Levi Van Sant, and Brian Williams. 2019. "Anthropo-cene, Capitolocene, . . . Plantationocene?: A Manifesto for Ecological Justice in an Age of Global Crises." *Geography Compass* 13 (5). https://doi-org.libezproxy2.syr.edu/10.1111/gec3.12438.

de la Cadena, Marisol, and Mario Blaser, eds. 2018. *A World of Many Worlds.* Durham, N.C.: Duke University Press.

Erev, Stephanie. 2019. "Feeling the Vibrations: On the Micropolitics of Climate Change." *Political Theory* 47 (6): 836–63. https://doi.org/10.1177/0090591719836195.

Escobar, Arturo. 2018. *Design for the Pluriverse: Radical Interdependence, Autonomy, and the Making of Worlds.* Durham, N.C.: Duke University Press.

FAO (Food and Agriculture Organization). 1999. "Background Paper 1: Agricultural Biodiversity." Presented at the Multifunctional Character of Agriculture and Land Conference, Maastricht, September. https://www.fao.org/3/x2775e/X2775E03.htm.

FAO (Food and Agriculture Organization). 2017. "6 Ways Indigenous Peoples are Helping the World Achieve #ZeroHunger." Food and Agriculture Organization of the United Nations, August 9. https://www.fao.org/indigenous-peoples/news-article/en/c/1029002/.

Figueroa-Helland, Leonardo, Cassidy Thomas, and Abigail Pérez Aguilera. 2018. "Decolonizing Food Systems: Food Sovereignty, Indigenous Revitalization, and Agroecology as Counter-Hegemonic Movements." *Perspectives on Global Development and Technology* 17 (1–2): 173–201. http://dx.doi.org/10.1163/15691497-12341473.

Gibson-Graham, J. K. 2006. *A Postcapitalist Politics.* Minneapolis: University of Minnesota Press.

Gilbert, Scott F. 2017. "Holobiont by Birth: Multilineage Individuals as the Concre-tion of Cooperative Processes." In *Arts of Living on a Damaged Planet: Ghosts and Monsters of the Anthropocene,* edited by Anna Lowenhaupt Tsing, Nils Bubandt, Elaine Gan, and Heather Swanson, M73–M89. Minneapolis: University of Min-nesota Press.

Haraway, Donna J. 1988. "Situated Knowledges: The Science Question in Feminism and the Privilege of Partial Perspectives." *Feminist Studies* 14 (3): 575–99. https://doi.org/10.2307/3178066.

Keeve, Christian B. "Fugitive Seeds." *Edge Effects,* February 5, 2020. https://edgeeffects.net/fugitive-seeds/.

Kimmerer, Robin Wall. 2013. *Braiding Sweetgrass: Indigenous Wisdom, Scientific Knowledge, and the Teachings of Plants.* Minneapolis: Milkweed.

Law, John. 2015. "What's Wrong with a One-World World?" *Distinktion: Scandina-vian Journal of Social Theory* 16 (1): 126–39. https://doi.org/10.1080/1600910X.2015.1020066.

Lax, Simon, Daniel P. Smith, Jarrad Hampton-Marcell, Sarah M. Owens, Kim M. Handley, Nicole M. Scott, Sean M. Gibbons, et al. 2014. "Longitudinal Analysis of Microbial Interaction Between Humans and the Indoor Environment." *Science* 345 (6200): 1048–52. https://www.science.org/doi/10.1126/science.1254529.

Nazarea, Virginia D. 2005. *Heirloom Seeds and Their Keepers: Marginality and Memory in the Conservation of Biological Diversity.* Tucson: University of Arizona Press.

Nazarea, Virginia D. 2021. "Ontologies of Return: Terms of Endearment and Entanglements." In *Movable Gardens: Itineraries and Sanctuaries of Memory,* edited by Virginia D. Nazarea and Terese V. Gagnon, 253–72. Tucson: University of Arizona Press.

Nazarea, Virginia D., and Terese V. Gagnon, eds. 2021. *Movable Gardens: Itineraries and Sanctuaries of Memory.* Tucson: University of Arizona Press.

Paxson, Heather. 2012. *The Life of Cheese: Crafting Food and Value in America.* Berkeley: University of California Press.

Ryan, John C., Patricia Vieira, and Monica Gagliano, eds. 2021. *The Mind of Plants: Narratives of Vegetal Intelligence.* Santa Fe, N.Mex.: Synergetic Press.

Scott, James C. 2009. *The Art of Not Being Governed: An Anarchist History of Upland Southeast Asia.* New Haven, Conn.: Yale University Press.

TallBear, Kim. 2019. "Caretaking Relations, Not American Dreaming." *Kalfou* 6 (1): 24–41. https://doi.org/10.15367/kf.v6i1.228.

Thomas, Deborah A. 2019. *Political Life in the Wake of the Plantation: Sovereignty, Witnessing, Repair.* Durham, N.C.: Duke University Press.

Tsing, Anna Lowenhaupt. 2017. "A Threat to Holocene Resurgence Is a Threat to Livability." In *The Anthropology of Sustainability: Beyond Development and Progress,* edited by Marc Brightman and Jerome Lewis, 51–65. London: Palgrave Macmillan.

White, Carol Wayne. 2017. "Stubborn Materiality: African-American Religious Naturalism and Becoming Our Humanity." In *Entangled Worlds: Science, Religion, and Materiality,* edited by Catherine Keller and Mary-Jane Rubenstein, 251–73. New York: Fordham University Press.

PART I

Refugia

SET THE GARDEN ON FIRE
    for Jeanette Li

My friend's new neighbors in the suburbs
are planting a neat row of roses
between her house & theirs.

Her neighbors smile, say the roses are part
of a community garden project, that's all.
But they whisper, too—whisper plans for trees,
a wall of them. They plant rumors
that her house is hiding illegals, when it's aunts
& uncles, visiting. They grow tall accusations
fed by talk radio, that her house was bought
with drug money, not seventeen years of woks
sizzling, people serving, delivering, filling,
people scrubbing, refilling, running—her family
running the best restaurant in town.
Like with your family, my friend says, *once we
moved in, they stopped calling us
hardworking immigrants.*
Friend, let's really move in, let's
plunge our hands into the soil.
Plant cilantro & strong tomatoes,
watermelon & honey-hearted cantaloupe,
good things, sweeter than any rose.
Let's build the community garden
that never was. Let's call the neighbors
out, call for an orchard, not a wall.
Trees with arms free, flaming
into apple, peach, pear—every imaginable,
edible fire.

Come friend, neighbor,
you, come set the garden on fire
with all our hard-earned years, tender labor
of being here, ceaseless & volcanic
making of being here, together.

                                        —*Chen Chen*

CHAPTER ONE

Epistemic Entanglements and Subsistence Corn Farming in Rural El Salvador

MIKE ANASTARIO

I recently had a flashback of refugia. The memory was intimate, painful, and brimming with agrobiodiversity. The vivid image was triggered when I heard an ophthalmologist say the word "scarring." She was gazing into my eyeball, and her assistant diligently wrote down my maladies. As a participant observer, I had, five years earlier, written my own field note after hacking into the soil with a sickle: "As I chopped at the dirt, some dirt flew in my eye. And it made me think almost immediately about what was in the dirt—were there any chemicals in it that I should be worried about?"

I remained silent as this memory passed through my mind, and my silence yielded the unintended consequence of no further investigation by the ophthalmologist despite the everyday hypothesis that was now spinning in my head: *Those agrichemicals did something to me.* Agrichemicals do a lot of things to a lot of bodies of people who rarely complain but frequently wonder about them. Agrichemicals contaminate soils that plants nonetheless find ways to grow in. From 2015 to 2020, I conducted fieldwork in and around Chalatenango, a rural northern region of El Salvador, where agrobiodiversity is found in small subsistence farms, and where farmers narrate and remember stories of agrichemicals entering the region. My flashback of refugia is not traumatic, because it is overpowered by the intimacy of cultivating *maíz capulin* (capulin corn) and *maíz blanco* (white corn) from soil that was and will continue to be contaminated with agrichemicals (figure 1.1). It was one fleeting reassembly of refugia, memory, and embodiment.

FIGURE 1.1 *Maíz capulin* (capulin corn) and *maíz blanco* (white corn) being placed into a pile on a milpa in Chalatenango. Photo by author.

This chapter is concerned with epistemic entanglements amid the everyday farming practices of subsistence corn farmers who live in, emigrate from, and send parcels to and from rural El Salvador. I first provide some context regarding the transnational and affective processes implicated in this agrarian assemblage, and then describe a minor intervention that slices, pokes, and pulls at imbricated knowledges therein. Findings from this small and nongeneralizable process concern soil chemistry in relation to local farming techniques, calling attention to epistemic entanglements in the material world of subsistence farmers, as well as the methods that social scientists rely on to examine them.

Corn and a Rural Salvadoran Diaspora

In the creation narrative of the Popol Vuh, grandmother Xmucane grinds white corn and yellow corn to make the flesh of the first humans. Corn is also the plant that Xmucane's grandchildren (Hunahpu and Xbalanque) use to provide a sign to their grandmother of their death and life before they embark on their journey to Xibalba (Tedlock 1996). While this narrative is almost never discussed among the subsistence farmers with whom I farmed corn in Chalatenango, its absence did not preclude our kneeling before every planted seed and manipulating the terrain to aid its growth in the twenty-first century.

In the second decade of this twenty-first century, the harvested corn will be brought to a Chalateca grandmother, who will compile *maíz blanco* seeds into a woven sack, which she will send to her migrant kin living in Colorado. She will send them by means of a courier, who will examine them for narcotics before packing them into a polypropylene-lined carboard box, which is one of multiple layers of protection for what will soon become airborne corn seeds. Eventually, the corn seeds will be stripped back down as an isolated parcel presented to a U.S. Customs and Border Protection agent in Houston who wants *to know what this is.* The corn seeds will be used to make atoll or nixtamalized tortillas in a place far away from the location in which the corn was grown. If I turn on a tape recorder and ask the parcel's recipient questions about why she requested the seeds, she might start by making comparisons between nation-states, people, sounds, and spaces. She might also perform nostalgia for my audio recorder. If in Chalatenango I ask her father to show me how he cultivates *maíz blanco,* he may choose to show me only parts of the process. He might secretly set fire to the weeds, producing an ash residue on the topsoil, after I leave because he knows that gringos do not like it when campesinos use fire to farm their food. He will smother his testimony simply because I am present, watching and observing (McKinnon 2016). The farmer and I might later talk until we are lost in wonder about what is really happening in the soil beneath our feet from which the *maíz blanco* grows. We might again kneel down on the soil before each seed in the coming rainy season, an act that clears the weeds but nonetheless causes us to genuflect before these signs of life and death.

Memory, Traditional Knowledge, and Inductive Learning

The traditional seeds, the farmer, the kin, the longing daughter consuming the corn in a faraway place, the observing and possibly judgmental gringo sociologist, and various knowledges are briefly entangled in the space and time of rural El Salvador. My focus on knowledge in this chapter is concerned with the complex and ongoing process of epistemicide (the killing of knowledge systems), the slow and unsurprising nature of which carries implications for the cultivation of traditional cultivars. I examine modern and traditional agricultural knowledge in this context. Both forms of knowledge are now entangled in collective memory, which is not something we have but something we do (Olick and Robbins 1998). Knowledge practices are impartially imbricated with memory practices, in part promulgating their existence and their effects.

Memory *practices* in the Salvadoran diaspora include particular styles and methods for connecting to and moving nonhuman actors, such as *maíz blanco* seeds, through the soil and transnational space. Today, Salvadorans in the United States represent approximately one-fifth of the total population of El Salvador (Terrazas 2010). Salvadoran diasporas have developed to the point that the Salvadoran government currently recognizes hundreds of *gestores de encomiendas* (parcel managers), or transnational couriers. Salvadoran couriers transport millions of U.S. dollars in cash and thousands of pounds of merchandise annually between El Salvador and the United States (Garni 2014; Anastario 2019). They collect parcels in El Salvador, travel to the United States, and deliver objects (and sometimes people) within diasporic networks. Food, medicine, plant products, documents, photographs, hammocks, clothes, school supplies, electronics, remittances, and letters are transmitted at high throughput across national airspaces to and from El Salvador.

In the Salvadoran diaspora, memory practices move parcels, and the movement of parcels stokes memories. Memory practices in the space of rural El Salvador also preserve traditional knowledge that may be used to grow corn in ways not recommended by state extension agents with expertise in modern agricultural science. Indeed, the repetitive use of fire on milpas is one such example of the irreverence shown by subsistence farmers to the hegemony of modern agricultural science (Nazarea 2005). This irreverence is marked by the entanglement of traditional fire and fallow methods imbricated with sequences of agrichemical applications on everyday subsistence farms in El Salvador.

The synchronization of memory practices and hybridized knowledge in a twenty-first century rural diaspora materializes in the movement of biodiverse actants like *maíz blanco*, with stalks and cobs that grow appreciably greener and larger depending on what farmers do during its growth. What farmers do on their own plots of land in the rural north often reflects inductive learning processes engaged in by attuning to the sequence of events that made those particular stalks and leaves grow greener, as their parents had, and as their grandparents had. It is not one (traditional knowledge) nor the other (modern agricultural science) type of knowledge that subsistence farmers in rural El Salvador rely on to inform the growth and regrowth of corn. The knowledge of what makes corn grow large and the leaves grow green is wound up in collective memory, knowledge regimes, and sensuously charged inductive learning. Attunements to avian migration, lunar cycles, and ongoing experimentation with technologies and timing generate iterations of life and death on local farms. Multiple forms of knowledge are imbricated with plant materiality, be it through inductively based attunement to one's environment, lessons from a parent, or knowledge disseminated by an extension agent. Focused on this materiality, the multiple elements and interactions that transform agrarian assemblages each year make prediction and repetition impossible (Coole and Frost 2010), despite appearing to repeat themselves as farmers attempt to predict plant growth. Inductive learning, in this context, is predicated on the accumulation of experience and updated decision-making strategies.

Epistemic Power and the Nonhuman

Contingencies between knowledge and materiality on farms can be examined relative to the hierarchies of power that occur between knowledge regimes. Knowledge produced and generated in the West, namely in Western universities, is often ascribed a "superiority" in comparison to the knowledge produced by alternative cosmologies and epistemologies that are divergent with, and often subject to colonization processes by, the West. This superiority is presumed by its practitioners. The universality of Western knowledge corresponds to an epistemic privilege that engenders epistemic inferiority among those outside and unsubscribing to the Western knowledge production complex (Grosfoguel 2013; Mignolo 2009). Western knowledge may be invoked and have the consequence of epistemicide—the killing of traditional (and often Indigenous) knowledge systems over long periods (Hall and Tandon 2017). Institutions train new members to be epistemically obedient while controlling who has access to narrow forms of knowledge generation (Mignolo 2009). Modern agricultural science provides an example of how scientific practice, technological developments and prescriptions, and extension agents are implicated in colonial and neocolonial processes of epistemicide that alter and erase the everyday knowledge, and corresponding farming practices, used to grow food. This erasure of traditional ecological knowledge is not a direct assault, but rather facilitated through processes such as hermeneutical marginalization (McKinnon 2016), in which the traditional knowledge of farmers is often denigrated, ignored, and stigmatized. Indeed, the "father" of the green revolution, Norman Borlaug, wrote that demonstration plots with "spectacular increases in yield destroy, in one stroke, the built-in conservatism or resistance to change that has been passed on from father to son for many generations in a system of traditional agriculture" (Borlaug et al. 1969). The endurance of traditional fire and fallow methods in rural communities throughout El Salvador, however, aggravates and interrupts the deployment of modern agricultural science by state extension agents. These aberrations create a space for hesitation (Lyons 2016) and invite us to consider questions that emerge among practitioners who know how to grow agrobiodiverse crops in refugia.

Knowledge regimes produce communities of practitioners in their wake (Strathern 2004), with consequences for plants, insects, the soil, and subsistence farmers touched by them. Traditional corn-farming practices in El Salvador are difficult to disentangle from practices that may have existed before the Spanish conquest. The word *milpa* itself is derived from the word for "field" in Indigenous Nahuatl and is today used to refer to an individual corn farm that may be intercropped with legumes, squash, and other plants (Harrison 2007). Worship of the water goddess, the sun, mother earth, mother moon, and plants prior to Spanish control has been described in local Indigenous cosmology and was evident in late nineteenth-century ethnographic observations (Hartman 2001; Cosgrove 2010), but these practices are highly eroded (approximating absent) following colonization. In nearby Guatemala, the milpa has been described as an agroforestry practice that employs perennial intercropping swidden cultivation techniques centered on corn, producing a diet that supplemented hunting, scavenging, and gathering practices as far back as the Archaic period (Ford and Nigh 2015). The contemporary Salvadoran milpa remains a site where the *three sisters* cultivation method is sometimes implemented (intercropping corn, legumes, and squash; Anastario, Salamanca, and Hawkins 2024).

The erosion of traditional farming knowledge in El Salvador was largely facilitated by the colonial eradication of Indigenous landholdings, potentiated by the genocide of Indigenous Salvadorans and subsequently buttressed by the formation of an agricultural extension agency. The archaeological site Joya de Cerén reveals that swidden *maíz*-milpa farming of the three sisters dates back to at least AD 600 in El Salvador (Sheets et al. 2011). After El Salvador's independence in 1824 and subsequent focus on agricultural development, tensions over land use intensified, culminating in privatization reforms that redistributed communal lands, or ejidos, to support entrepreneurs (primarily in coffee; Browning 1987). Legal steps to dismantle communal land ownership commenced in 1880, as the government pushed for privatization to increase agricultural efficiency and exert social control by turning dispossessed peasants into a labor force (Browning 1987; Vega Trejo et al. 2019). In the early 1880s, the Salvadoran government dissolved Indigenous communal landholdings that would have more readily supported traditional agricultural practices in these communities (Vega Trejo et al. 2019), a move that was interpreted as a symbol of modernization and prosperity. The genocide of thousands of Indigenous and Ladino peasants during a military-led coup in the early twentieth century (Gould and Lauria-Santiago 2008; Ching 2014), and the apparent absence of indigenously identified Chalatecos in the networks studied, has substantially diminished Indigenous Salvadoran knowledge as it relates to milpa-farming practices. In 1942, the secretary of agriculture and the U.S. Department of Agriculture signed a memorandum of understanding to establish the Centro Nacional de Agronomía (National

Agronomy Center), which in the twenty-first century would be renamed the Centro Nacional de Tecnología Agropecuaria y Forestal (National Center of Farming and Forestry Technology, or CENTA), to represent the government's agricultural extension efforts, imparting the lessons of modern agricultural science throughout the Salvadoran countryside (CENTA "Enrique Álvarez Córdova" 2015).

These developments corresponded to the implementation of the green revolution, during the 1960s and 1970s, in which selection, breeding, culling, and technological development improved yields but shifted reliance away from traditional knowledge and cultivars (Nazarea 1998). In El Salvador, the implementation of the green revolution is temporally associated with dramatic increases in yields, with bean and corn production quadrupling from the 1950s into the second decade of the twenty-first century (Flores Romero and Orellana Guevara 2019).

Given these historical inputs, note that Chalatenango (the rural north) is where economically poor campesinos have historically had more control over their land and more freedom to develop their own paths than their counterparts in the larger coffee-export zones of El Salvador, which were dominated by members of the elite class (Todd 2010). This independence has in part contributed to the syncretism in modern and traditional farming practices that can be observed today on small subsistence farms in Chalatenango. Given the genocides and epistemicide in El Salvador, "traditional knowledge" is representative not of Indigenous knowledge regimes in El Salvador prior to the Spanish conquest, but instead practices that remain transmitted orally and that are learned inductively by farmers.

Re-turning Epistemic Relations in El Salvador

Farming practices that manipulate plants, soil, and ecosystems are imbricated with epistemic superiority, inferiority, and inductive learning. My imperfect but active methodological experiment begins at this juncture. I aimed to work with local subsistence farmers to intervene in and re-turn knowledge practices that are entangled with Salvadoran subsistence farms, altering the act of participant observation itself. Karen Barad writes that *re-turning* does not refer to "returning as in reflecting on or going back to a past that was, but re-turning as in turning it over and over again—iteratively intra-acting, re-diffracting, diffracting anew" (Barad 2014, 168). I use Barad's notion of "re-turning" the past, as it can indeed trouble binaries and clean lines of demarcation that inform knowing, opening up new ways of understanding the material world and the agencies in it. More specifically, by pulling apart some knowledge practices imbricated with corn growth that I would have otherwise "observed," I was able to better understand how the components of the entanglement snap back together.

Key informants led me to others, and as the rainy season neared, our conversations began to focus on the probable timing of the first rains. Farmers used not books but conversation that encoded how the lunar cycle, weather patterns, and avian migration temporalities indicate the best moments for sowing corn into the ground. It was nothing I could schedule in my planner, and it varied day by day. During this period, farmers also talked about what *plants wanted* and whether *god* would allow the earth to give. My gradual attunement to these rhythms, signals, and words gradually altered the human and nonhuman networks I interacted with, as I increasingly became connected to farmers who used nonhuman signals and actants to grow their corn. Traditional knowledge embedded in collective memory practices, such as storytelling, emerged as the rainy season neared. One campesino farmer now in his sixties remembered a time during the dry season when

> we were just kids, and [dad] told us to "prepare the spades because tomorrow we are going to sow." And my brother, who is now living in Colorado[,] says "fuck, dad is crazy." My brother said to him "but how are we going to sow if the ground is so hard and it hasn't rained?" Dad told him "prepare the spades because tomorrow we are going to sow." We went along with my father and we sowed corn in total disbelief. And that afternoon, the storm fell and the rainy season began. . . . I don't know, they had that—as one would say—that wisdom.

My interest in subsistence farming and heirloom seed preservation further drove my networking with farmers who preserve their own heirloom seeds, and one family gave me several pounds of corn, bean, and squash seeds to cultivate. None of the seeds was purchased in a store, and the seeds generally met the heirloom criteria (Nazarea 2005). Since I had the privilege of cultivating for interest as opposed to subsistence, I used this opportunity to intervene in participant observation. Here, I introduced the rule that we could not grow corn with pesticides (based on local farmers' memories), which is a type of temporal-technological diffraction in this entanglement. Further, I acknowledge my intervention into the observer/observed boundary by asking the farmers to help me maintain experiments that they, by default, designed as farming moved forward.

First, farming a milpa without using herbicides or pesticides requires that farmers clear the land by *pura cuma* (pure sickle), which is physically laborious. It is also a style of clearing weeds that was practiced prior to the introduction of agrichemicals in the region, stimulating memories of farming (not only with those who helped me work on the milpa, but with those community members who passed through the milpa). We would spend hours kneeling on the ground before each planted corn seed, using the sickle to hack circles into the soil surrounding each planted seed, building up the soil around the seed while clearing weeds from where it would grow. It was difficult but meditative work, which often provided the space for long conversations to emerge and sustain amid the rhythm of repetitious genuflection. In return for the heirloom seeds, I paid the farmers for their labor when I needed help with farming tasks on the land that was lent to me. I had observed subsistence farmers paying other farmers for labor on their own milpas the year prior, and a key informant determined the going rate.

Second, disrupting the observer/observed boundary required that experimental design and nonhuman observation be shared by farmers who were not necessarily interested in conducting a study from the outset. As such, the design evolved as farming moved forward. There were no a priori controls. The farmers that I was working with used fire to burn organic waste in two locations on the milpa, and they did so on their own terrain as well. As a memory practice, the use of fire on the milpa was connected to memories of the past. The practice is widespread among farmers in Chalatenango, but it contributes to epistemic injustice (McKinnon 2016), in that farmers' persistent use of fire is characterized by agronomy experts, community members, and outsiders as problematic, backward, uneducated, and contributing to environmental degradation. Aside from the fire used on our plot, one of the participating farmers transmitted a video of his brothers burning his own milpa with the voiceover "allá están los bichos ya dándole fuego ahí al monte" (there are the guys setting the weeds on fire).

In their own personal practice on their own land, the subsistence farmers I worked with also combined firewood ash with synthetic fertilizers (ammonium sulfate and 15% nitrogen [N] / 15% phosphorous [P] / 15% potassium

[K]) for the second round of corn fertilization, which generally occurs four weeks after sowing corn. Another farmer who used this technique explained to me that "la ceniza es vitamina para el maíz" (ash is vitamin for corn). He acknowledged the value of this knowledge by remembering how his own father, back in the 1950s (prior to the green revolution), used to bring him to the farm as a child and how "cortábamos el monte, quemábamos, y sembrábamos. La milpa da buena cosecha porque hay materia podrida y queda mucha ceniza. Solo se hacía dos veces. Después, se descansa la tierra" (we used to cut the weeds, we burned them, and then we planted [corn]. The milpa gives a good harvest because there's decaying material and a lot of ash. It was only done two times. Afterward, the land rests.)

Older farmers who remembered pre–green revolution farming practices described two-year land use cycles before clearing (*rozar*) new milpa space. Farmers with pre–green revolution memories remembered cultivating cornfields with a topsoil layer of ash over six inches thick. I listened to farmers imagine the properties of the ash as fertilizer, describing how perhaps the ash washed some type of *lejía* (lye) into the soil that could explain why the ash caused corn to grow as it did when farmers were doing something they were otherwise pressured not to do (swidden farming). Despite the shaming and scolding that farmers have since received for using fire on their milpas, subsistence farmers continued to use the traditional knowledge of swidden farming to grow their corn in 2018.

In the post–green revolution era, when agricultural knowledge is managed by agents of the state, swidden techniques are often discouraged by extension agents and stigmatized by community members. The logic is that swidden farming contributes to erosion, hydrophobic topsoil, and long-term nutrient depletion. What was once a technique for pest control, fertilization, and land management was, at the time I was conducting my field observations, a technique referenced by others to stigmatize farmers, who anticipate the stigma and hide their use of fire to the best of their ability. Swidden techniques have alternatively been described as creating pastures and stimulating forest heterogeneity and biodiversity (Tsing 2015; Eckwall 1955; Siebert and Belsky 2014; Padoch and Pinedo-Vasquez 2010). This human-generated disturbance created micro-patches on the milpa shaped by fire conjunctures (a nonliving disturbance driven by humans; Tsing 2015). The swidden farming practices were partially derived from practices of the past and inductively learned over time.

Intervening into the observer/observed dichotomy, I subsequently de-marcated five patches that the farmers had designed on the milpa following the rule of no herbicide or pesticide use, each with their own dose of nonliv-ing disturbance created by humans, which had emerged in situ:

1. A swidden plot (where organic waste was set on fire, leaving an ash residue on the topsoil) where no synthetic fertilizers would be added
2. A swidden plot where synthetic fertilizers would be added
3. A no-burn plot where no synthetic fertilizers would be added (the control plot)
4. A no-burn plot where only synthetic fertilizers would be added
5. Intervention #4 with a fifteen-day delay in fertilization timing

The farmers could readily recognize these conditions, as they generated them and inherently understood how they demarcated the different zones (figure 1.2). I had already conducted soil chemistry analyses at baseline (35 days prior to sowing corn), and I collected follow-up samples at 105 days following corn being sowed (which corresponded to the period in which the stalks were folded), or 140 days following the baseline soil sample collection.

This type of experiment produced a highly local type of statistically in-significant data relevant to local farming practices. It evaluated everyday hy-potheses that had emerged amid refugia, even if those hypotheses, practices, and my evaluation of them were illegible to Western knowledge (Escobar 2016), while drawing methods from Western science to examine what had emerged in the micro-patches. Baseline soil plot samples were pooled to create baseline soil chemistry values and then compared to postintervention emergent patches.

The quasi-experimental and flawed design enabled different intervention types to emerge at the same time, allowing for more power sharing in terms of the defined intervention and the implementation of the intervention. It also helped me explore questions that I had heard farmers talking about the year prior—namely, what is going on in the soil relative to the techniques they used, when mixtures of fertilizers and/or ash are added to and taken from milpa ecosystems. The farmers who worked with me managed the plots of their own making in size and technique, given my external rule of no pesticide use, and defined their own experimental condition (plot 5) based

FIGURE 1.2 Different zones within the milpa (picture taken from center of milpa).
Photo by author.

on their assumption that synthetic fertilizer administration at too early a stage of corn growth would affect the vertical growth of corn (which was later "validated" at 140 days). CENTA's soil chemistry laboratory analyzed the samples (CENTA 2018), blinded from what we were doing.

As we farmed the site, older farmers often stopped by and asked why we were not using pesticides to clear the weeds. Several volunteered to help us for short stints of time and even noted while cutting weeds with the sickle that this was what it was like to farm during childhoods long ago. There was also gossip circulating in town that I was letting my milpa become overgrown with weeds, and that some of my corn (the control plot) was yellowing and in need of synthetic fertilizer. On the control plots, the small or absent corncobs growing on the plant were a point of almost public shame, as numerous farmers who passed by on a footpath through the milpa could observe and comment on what was happening on the farm. Had I no idea how to grow corn, the enacted stigma from other farmers would have quickly pointed me toward numerous agrichemicals available at local stores throughout town. Some older farmers even arrived to the milpa as I was farming to ask why I had not applied paraquat or N/P/K fertilizer combinations on the terrain.

While synthetic fertilization practices vary by farmer, we followed the general fertilization guidelines provided by CENTA in micro-patch conditions designated for synthetic fertilizer application. CENTA guidelines for corn fertilization given baseline soil chemistry analysis included liming one month prior to planting (the soil pH was 5.43 at baseline); 15%N/15%P/15%K, zinc, and copper at sowing; ammonium nitrate at four weeks postplanting; and urea at seven weeks postplanting. We did not follow liming, zinc, nor copper recommendations because of timing and logistic constraints. Further, the ammonium nitrate that was recommended could not be found commercially, and the widely available ammonium sulfate was instead applied. In the interest of consolidating the presentation of results for this component of our research, I report findings for the following nonhuman variables within each micro-patch on the milpa at baseline and follow-up: pH in water, phosphorous (P), potassium (K), calcium (Ca), magnesium (Mg), sodium (Na), cation-exchange capacity (CIC), and percentage base saturation. I calculated the raw effect size (ES) for the emergent micro-patch such that any given $ES = (I_{t2} - B_{t1}) - (C_{t2} - B_{t1})$, where I is the intervention value at $t_2 = 140$ days following the initial soil test, B is the baseline value at $t_1 = 0$ days, and C is the control group value. I report the unstandardized ES in the text below, and for a single visual representation of nonhuman actors in the soil, I standardized chemical values ($z = [x - \mu]/\sigma$) prior to calculating the ES for select nonhuman chemical actors presented in figure 1.3. While results are based on a considerably small sample size ($n = 5$), the purpose of this analysis is not to generalize but to see, complicate, and unpack the materiality of epistemic entanglements on a milpa.

At the 140-day follow-up, the swidden patches were the only zones where pH increased and where base saturation reached 100 percent (both had an ES of +1.4), compared to the base saturation ES for synthetically fertilized patches without ash fertilization (−5.8 and −9.3). The ES for cation-exchange capacity was +1.3 in the swidden plot without synthetic fertilizer, and +5.9 in the swidden plot with synthetic fertilizer, with unsubstantial ES (<0.3) observed in patches without ash fertilization. For percentage organic material, the swidden micro-patch with no synthetic fertilizer showed the largest ES (+0.68), followed by the micro-patch where swidden and synthetic fer-

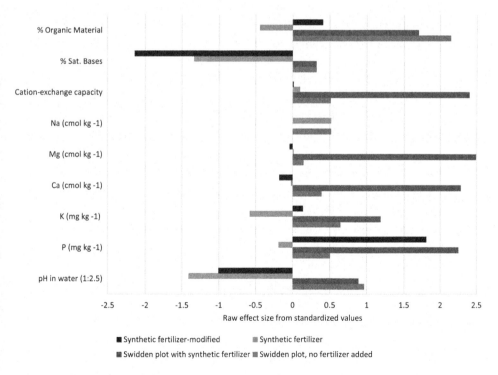

FIGURE 1.3 Standardized effect sizes for soil chemical properties relative to patches of fertilization on the milpa. Credit: Mike Anastario

tilization techniques were combined (+0.54). A visual summary of the soil chemistry results is provided in figure 1.3, where medium-gray bars represent patches where fire was used.

The more favorable soil chemistry results corresponded to the micro-patch denoted as the swidden plot with synthetic fertilizer. This micro-patch condition was most similar to the farmers' traditional practices of swidden agriculture, which included mixing ash with synthetic fertilizers at four weeks postsowing. This micro-patch showed the largest effect sizes for calcium, magnesium, potassium, and CIC in comparison to all other patches. Further, these patches were associated with visually apparent plant growth and cob size relative to growth on the control plot.

In contrast to the prevalent use of fire and ash to grow corn among the network of subsistence farmers I was connected to in Chalatenango, the relative dismissal of swidden farming by noncampesinos with "more education" was appreciable. One agronomic expert at CENTA grew frustrated with me as I explained this experiment, and he interrupted my description of the soil tests to inform me that the soil in the micro-patch where fire was used had been ruined (without seeing the soil chemistry results). Further, he interpreted my presence as a sociologist to suggest that I should be conducting a knowledge, attitudes, and practices survey that could help extension agents deploy the appropriate knowledge to this particular zone. In other words, a type of agricultural knowledge that was locally valued and repeatedly practiced on the milpa in Chalatenango was readily dismissed by a knowledge manager who knew so much that he did not even need to let me describe what we did, or why we did it. Despite my privilege, this automatic dismissal, so minor but so significant, is one of many elements that arise when I tug at and question the knowledge firmly entangled with plants, soils, and farmers in twenty-first-century Chalatenango. The dismissal is an everyday act of epistemic violence, which contributes to the epistemic oppression of Chalateco farmers, many of whom engage in "testimonial smothering" (McKinnon 2016) as they anticipate similar hostilities from educated experts who do not live near them nor farm their lands. In other words, undermining modern agricultural knowledge resulted in an expert snapping at people who seemed to be irreverent to its injunctions, thereby elucidating one of the ways in which the epistemic entanglement remains tangled. Subsistence farmers at the margins will continue to practice techniques that are irreverent to the hegemony of modern agricultural science, often in "secret," despite the enacted stigma, drawing from what is remembered and inductively learned over time.

Conclusion

I open this chapter with a brief vignette about memory, refugia, and embodiment. Everyday hypotheses are often formulated but never expressed by farmers who, despite what so many people say about and to them, seem to have a way of getting traditional cultivars to grow in the most unlikely spaces. I illustrate an experiment that permitted a space for farmers to explore some everyday hypotheses, and the experience elucidated how this type of endeavor could be perceived as epistemically disobedient, and how social processes buttress the testimonial smothering of those who know how to grow traditional corn at the rural margins of El Salvador (Mignolo 2009; McKinnon 2016).

The word that Chalateco subsistence farmers use to describe returning to the same milpa plot that had been cleared with fire and farmed the year prior is *huatal*, which is similar to the Nahuatl *ouatl* and used to describe an area with crop residues and grasses that had been burned and cultivated during the previous rainy season (Jansen 1998). Traditional swidden farming knowledge, evident in memory practice, is now subject to epistemicide, even if that epistemicide is unintentional, by agronomy experts and even scholars who discourage the use of fire. Indigenous farming practices in places like Chalatenango, where less than 1 percent of Chalatecos today claim Indigenous identity (Lenca, Kakawira, Nahua-Pipil, or Other), now exist in a syncretic form among farmers who do not claim Indigenous identity, nor can even recall memories of Indigenous peoples in Chalatenango, but who use Indigenous techniques in farming practices they remember using for quite some time. When I tugged at epistemic entanglements in this agrarian assemblage, a brief encounter with epistemic violence showed me how it snaps back into place.

While the findings in this chapter do little to alter the rulings of agronomic experts and extension agents who deter the use of fire and deploy modern agricultural science in the countryside, the results simultaneously illustrate why subsistence farmers might perceive benefits from using fire to grow corn. Further, the amendment to participant observation permitted a degree of control in the experiment by farmers who are not just familiar with *but experts in* subsistence farming practices, despite their testimonial smothering. The lead farmer who helped me with the experiment explained how he had learned something from the endeavor. What was unique for him about the experience was the ability to try five things at once without risking starvation, which was an opportunity that my (largely economic) privilege could afford in the context of participant observation.

Perhaps questioning knowledge by intervening in its perceived superiority can reveal the ways in which it materializes, benefits, and harms the numerous actors implicated in its entanglements. With the passage of time away from the field, I often spend time wondering what that field did to my eyes, my blood, and my body. I often wonder what farmers know but rarely say as they get traditional cultivars to grow again and again. The methods for making those seeds materialize and rematerialize in the world are so irreverent and effective, pulling from the dominant regime when desirable, concealing themselves at the margins, and promulgating their ongoingness through oral histories and inductive learning.

References

Anastario, Mike. 2019. *Parcels: Memories of Salvadoran Migration.* New Brunswick, N.J.: Rutgers University Press.

Anastario, Mike, Elena Salamanca, and Elizabeth Hawkins. 2024. *Kneeling Before Corn: Recuperating More-than-Human Intimacies on the Salvadoran Milpa.* Tucson: University of Arizona Press.

Barad, Karen. 2014. "Diffracting Diffraction: Cutting Together-Apart." *Parallax* 20 (3): 168–87. https://doi.org/10.1080/13534645.2014.927623.

Borlaug, Norman E., Oddvar Aresvik, Ignacio Narvaez, and R. Glenn Anderson. 1969. "A Green Revolution Yields a Golden Harvest." *Columbia Journal of World Business* 4:9–19.

Browning, David. 1987. *El Salvador: La Tierra y el Hombre.* 3rd ed. San Salvador: Dirección de Publicaciones.

CENTA (Centro Nacional de Tecnología Agropecuaria y Forestal). 2018. *Laboratorio de suelos.* Ciudad Arce, El Salvador: Ministerio de agricultura y ganadería.

CENTA (Centro Nacional de Tecnología Agropecuaria y Forestal) "Enrique Álvarez Córdova." 2015. "El centa: Evolución histórica y aportes al desarrollo agropecuario." Ciudad Arce, El Salvador: Gobierno de El Salvador.

Ching, Erik. 2014. *Authoritarian El Salvador: Politics and the Origins of the Military Regimes, 1880–1940.* Notre Dame, Ind.: Notre Dame Press.

Coole, Diana, and Samantha Frost. 2010. "Introducing the New Materialisms." In *New Materialisms: Ontology, Agency, and Politics,* edited by Diana Coole and Samantha Frost, 1–44. Durham, N.C.: Duke University Press.

Cosgrove, Serena. 2010. *Leadership from the Margins: Women and Civil Society Organizations in Argentina, Chile, and El Salvador.* New Brunswick, N.J.: Rutgers University Press.

Eckwall, E. 1955. "Slash and Burn Cultivation: A Contribution to the Anthropological Terminology." *Man* 55: 135–36.

Escobar, Arturo. 2016. "Thinking-Feeling with the Earth: Territorial Struggles and the Ontological Dimension of the Epistemologies of the South." *Revista de Antropología Iberoamericana* 11 (1): 11–32.

Flores Romero, R. A., and L. A. Orellana Guevara. 2019. "Modelos autorregresivos integrado de medias móviles (ARIMA) y vectores autorregresivos (VAR) aplicados a la producción nacional de maíz y frijol en El Salvador para el periodo 1955–2030." Tesis de la Maestría en Estadística Aplicada a la Investigación, Universidad Centroamericana José Simeón Cañas, Antiguo Cuscatlán, El Salvador.

Ford, Anabel, and Ronald Nigh. 2015. *The Maya Forest Garden: Eight Millennia of Sustainable Cultivation of the Tropical Woodlands.* Walnut Creek, Calif.: Left Coast Press.

Garni, Alisa. 2014. "Transnational Traders: El Salvador's Women Couriers in Historical Perspective." *Sociological Forum* 29 (1): 165–88. https://doi.org/10.1111/socf .12074.

Gould, Jeffrey L., and Aldo Lauria-Santiago. 2008. *To Rise in Darkness: Revolution, Repression and Memory in El Salvador, 1920–1932.* Durham, N.C.: Duke University Press.

Grosfoguel, Ramón. 2013. "The Structure of Knowledge in Westernized Universities: Epistemic Racism / Sexism and the Four Genocides / Epistemicides of the Long 16th Century." *Human Architecture: Journal of the Sociology of Self-Knowledge* 11 (10): 73–90.

Hall, B. L., and R. Tandon. 2017. "Decolonization of Knowledge, Epistemicide, Participatory Research and Higher Education." *Research for All* 1 (1): 6–19. https:// doi.org/10.18546/RFA.01.1.02.

Harrison, Peter. 2007. "Maya Agriculture." In *Maya: Divine Kings of the Rain Forest*, ed. Nikolai Grube, 71. Königswinter: H. F. Ullman.

Hartman, Carl V. 2001. "Reconocimiento Etnográfico de los Aztecas de El Salvador." *Mesoamérica* 22 (41): 146–91.

Jansen, Kees. 1998. *Political Ecology, Mountain Agriculture, and Knowledge in Honduras.* Amsterdam: Thela.

Lyons, Kristina. 2016. "Decomposition as Life Politics: Soils, Selva, and Small Farmers under the Gun of the U.S.-Colombian War on Drugs." *Cultural Anthropology* 31 (1): 56–81. https://doi.org/10.14506/ca31.1.04.

McKinnon, Rachel. 2016. "Epistemic Injustice." *Philosophy Compass* 11 (8): 437–46. https://doi.org/10.1111/phc3.12336.

Mignolo, Walter D. 2009. "Epistemic Disobedience, Independent Thought and Decolonial Freedom." *Theory, Culture, and Society* 26 (7–8): 159–81.

Nazarea, Virginia. 1998. *Cultural Memory and Biodiversity*. Tucson: University of Arizona Press.

Nazarea, Virginia. 2005. *Heirloom Seeds and Their Keepers: Marginality and Memory in the Conservation of Biological Diversity*. Tucson: University of Arizona Press.

Olick, J., and J. Robbins. 1998. "Social Memory Studies: From 'Collective Memory' to the Historical Sociology of Mnemonic Practices." *Annual Review of Sociology* 24:105–40.

Padoch, C., and M. Pinedo-Vasquez. 2010. "Saving Slash and Burn to Save Biodiversity." *Biotropica* 42:550–52.

Sheets, P., C. Dixon, M. Guerra, and A. Blanford. 2011. "Manioc Cultivation at Ceren, El Salvador: Occasional Kitchen Garden Plant or Staple Crop?" *Ancient Mesoamerica* 22 (1): 1–11.

Siebert, S. F., and J. M. Belsky. 2014. "Historic Livelihoods and Land Uses as Ecological Disturbances and Their Role in Enhancing Biodiversity: An Example from Bhutan." *Biological Conservation* 177:82–89. https://doi.org/10.1016/j.biocon.2014.06.015.

Strathern, Marilyn. 2004. *Commons and Borderlands: Working Papers on Interdisciplinarity, Accountability and the Flow of Knowledge*. Oxon: Sean King.

Tedlock, Dennis. 1996. *Popol Vuh: The Definitive Edition of the Mayan Book of the Dawn of Life and the Glories of Gods and Kings*. New York: Simon and Schuster.

Terrazas, A. 2010. "Salvadoran Immigrants in the United States in 2008." Migration Information Source, January 5. http://www.migrationpolicy.org/article/salvadoran-immigrants-united-states-2008.

Todd, Molly. 2010. *Beyond Displacement: Campesinos, Refugees, and Collective Action in the Salvadoran Civil War*. Madison: University of Wisconsin Press.

Tsing, Anna Lowenhaupt. 2015. *The Mushroom at the End of the World: On the Possibility of Life in Capitalist Ruins*. Princeton, N.J.: Princeton University Press.

Vega Trejo, Ana L., Kevin Rivera, Elena Salamanca, Donovan Najarro, Carlos Aguiluz, Lissette de Schilling, and Nelson Crisóstomo. 2019. *45 Años del Banco de Fomento Agropecuario Entretejiendo Vidas e Historia*. San Salvador: Banco de Fomento Agropecuario.

CHAPTER TWO

Plot Life in Flower City

Transnational Ritual Ecologies in the Wake of Plantations

SHANNON A. NOVAK

It was so hot. Incense and cigarette smoke hung in the air, stinging my eyes and lungs. Soot was drawn to the sticky residue on my skin, a mixture of sweat and burning camphor and ghee. Black rings were, no doubt, forming around my nostrils. A good puja could always be marked by the number of days I blew black snot out of my nose. Only four hours in on a sizzling July afternoon, and things were just getting ramped up. The "sweet work" was done, offerings of confections, fruits, and flowers made to the more ortho- dox Hindu gods—Ganesh, Krishna, Hanuman, among others—before turn- ing to those with different appetites. Cigarettes, rum, and the blood of fowl and goats are required to appease the darker forms. "Life work," or animal sacrifice, is performed off-site. Compromise must be made when inviting the goddess Mariamma to come sit within the city limits of Brampton, Ontario.[1]

How to begin the story of this weekly invitation that involves laying the table for her visit? And perhaps a better question, *where* to begin? Maybe South India, "home" of the goddess whose powers of fertility are invoked to bring rains to barren fields, babies to troubled wombs, and remove evil and other bodily afflictions (Biardeau 1989; Bloomer 2018; Trawick 2017). Or maybe British Guiana, the "new homeland" of indentured laborers who carried her across dark waters in the nineteenth century to the edge of cane fields on sugar estates (Bahadur 2014; McNeal 2011; Stephanides and Singh 2000; van der Veer and Vertovec 1991; Younger 2009). Following that na- tion's independence in 1966, and the economic and political upheaval that ensued, Indo-Guyanese emigrants took flight, bringing revitalized traditions from the Caribbean to North American cities (Hirsch 2019; S. L. Jackson 2016; Kloβ 2016; Thomases and Reich 2019). Traversing these generations and geographies is relevant to understanding how the goddess in her many forms has come to be revered in a barren industrial park on the outskirts of Toronto.

Brampton, rather ironically, has come to be known as Flower City as a result of the floriculture industry that once thrived there. The enterprise was initiated in 1863 by Edward Dale, an English immigrant to the rural village, who began by selling vegetables grown in his homegarden. For pleasure he tended roses on the side, something his son also developed a love for and used to charm women customers with as a gift (O'Hara 2007). Dale's success drove expansion of the business into greenhouse cultivation, first of vegetable varieties then exclusively roses. By the turn of the century, the nursery market had become the primary employer in the area, with acres of land cleared and devoted to growing flowers under glass. The nursery was renowned for its award-winning hybrids at international shows, and at the peak of production in the 1960s, workers cut 20 million blooms annually to be dispersed globally (O'Hara 2007).

Such a place, it seems, would have been amenable to Mariamma worship, given its roots in rural landscapes and the multispecies entanglements that entails. Even with the shift to an industrial scale, the ritual practice has shown it can persist at the plantation's edge. It would be decades, however, before the goddess would settle in Brampton in any institutionalized way. Still continents apart, these histories and fates were intimately entwined with nebulous global forces. Shockwaves from the 1970s oil crisis would cause both the Dale Estate in Brampton and the sugar estates in Guyana to collapse, bringing refugees of these two worlds together. But beyond a few road names, historical plaques, and city branding, little remains of Brampton's flower works today. Warehouses have replaced greenhouses, and meters of paved roads suffocate most signs of life. How does anything survive, let alone thrive, in the ruins of capitalism? And yet a vast transnational network of commodities and exchange has vitalized Mother worship here, even nourished its growth. This weedy tradition appears quite adept at colonizing cracks in the pavement.

How to begin, *where* to begin, indeed. With multiple itineraries that do not easily line up, creeping change and dramatic rupture, this entangled bank of the Anthropocene might best be approached as fertile "demonic ground"

(Davis et al. 2019; McKittrick 2006; Wynter and McKittrick 2015). Such an approach draws on Katherine McKittrick's disruptive and "undisciplined" model of placemaking, which edits cartographies of empire. By stepping out of—across, within, and between—naturalized borders and boundaries, as well as the teleological and temporal dimensions these imply, McKittrick (2006) makes visible Black women's lives and the diasporic terrains their presence cocreates.

Like McKittrick, my study draws on landscapes and creative placemaking in the Caribbean and Canada, with an added dash of India and other nation-states unnamed that feed into ritual supply chains. At the same time, my intention is not to conflate the histories and experiences of enslavement and indenture, but neither is it to keep them apart. To do so would merely replicate categories of empire as hard, inalienable facts used to carve up peoples and their place (Khan 2015; Lowe 2011; McKittrick 2021; Palmié 2023). Rather, by opening pathways and itineraries, radical solidarities (however fleeting) might appear during humble moments when histories and entities touch. What better place to begin than the shared dirt beneath the feet, and the uneven terrain of living with/in the plantation's wake.

The logic of these colonial-era systems is argued to reverberate so widely in the world today that little if anything remains untouched (e.g., Hartman 2002; Le Petitcorps, Macedo, and Peano 2023; McKittrick 2013; Rodney 1972; Thomas 2019; Trouillot 2002; Woods 1998). Variations on the plantation system persist—from "enclosures and reserves; industrial estates and mill villages; free-trade and export zones; enterprise and empowerment zones; ghettos and gated communities; suburbanization and gentrification; game preserves and tourist resorts; pine plantations and mines; and migratory and prison labor" (Woods 2007, 56). What unites these diverse settings is a formula for expansion that European planters "stumbled on" in the sixteenth and seventeenth centuries: "exterminate local people and plants; prepare now-empty, unclaimed land; and bring in exotic and isolated labor and crops for production" (Tsing 2015, 39). The destruction has been so successful that the "Plantationocene," rather than the more popular moniker of the Anthropocene, has been proposed as a more appropriate descriptor for the current global socioecological crisis (Haraway et al. 2015).

Importantly, these destructive legacies were not—are not—thorough or complete but spawned creative "acts of survival" (McKittrick 2013). From archives and images, Judith Carney traces some of these sites where biodiversity flourished: dooryard gardens and small farm plots in marginal terrain deemed unfit for cultivation. "From a contemporary vantage point, slaves' modest subsistence plots can be seen as islands of agrobiodiversity disrupting a sea of commodity monoculture" (Carney 2020, 1082). Dooryard gardens were particularly diverse, harboring a vast assemblage of flowers, fruits, plants, and trees that came from Old World and New. These "botanical gardens of the Atlantic world's dispossessed" (Carney and Rosomoff 2009, 135) contained not only an impressive inventory of species taxonomies, but knowledge and practices not so easily cataloged and named. Intimately positioned and tended daily, garden plots supplemented meager rations and soothed body and soul. Through an ethos of "good use" and reciprocity, the plot offered material and spiritual nourishment in ways the cane fields could not (Davis et al. 2019, 8). Along with a modicum of "spatial sovereignty" (Fellows and Delle 2015), the plot was a place to feel alive—to feel human—in a system intent on stripping some people of this status (Z. I. Jackson 2020; Mbembe 2019; Weheliye 2014).

The plot-and-plantation thesis that Sylvia Wynter initially laid out, and that McKittrick, Carney, and other scholars of Black ecologies build on, encapsulates "the basic confrontation" between these two landforms and "the structure of values which each represent" (Wynter 1971, 99). Importantly, these are not discrete or purely oppositional; rather, they are relational and lived. It is through movement within and between, to places elsewhere, where difference is felt. In the wake of the plantation, Deborah Thomas (2019, 1) asks us to ponder, "What does modern sovereignty *feel* like?" Plot living might shed some light. Such inquiries will require getting closer to the ground—down in the weeds, so to speak—where tenacious beings still flourish. Even more powerful, then, might be the counter exposed "in marginal fields and uncaptured spirits" (Nazarea 2006, 320). Such a place can be found in one of Brampton's many industrial parks, where the plot and plantation continues to reverberate.

Near the airport, off the expressway—no place really—take a left just before the train tracks into one of these parks. On your right, just past the roti shop, is a small oasis—though difficult to tell from outside. Within Unit 28, the goddess sits (figure 2.1). It is a refuge of hope and healing, of need and desire—a place where an aesthetic of beauty dominates, but it comes at a price. To ignite the many senses that overwhelm bodies during puja requires assembling with abundance. Rather than resist the market and blight that

FIGURE 2.1 Murti of Mariamma on the main altar at the Shri Maha Kali Amma Mandir, Brampton, Ontario. Note the two karagams that bracket Mother's feet, marked by their pyramid structure of oleander and neem, and the molded face of turmeric paste. Photo by author.

surrounds them, the devotees are imbricated in complex networks of exchange. These are not easily divided into moral or market economies (Graeber 2011), though balance sheets are kept. Like the commercial mushroom collectors Anna Tsing follows, "there is too much contingency and variation here to imagine a simple calculus of supply and demand" (2012, 152).

Some materials and substances store well at the temple—dry goods rest on shelves in backrooms and tucked beneath altar skirts. Others are more perishable and scheduled for weekly delivery at the back door, especially the imported "fresh" fruits, plants, and flowers that take center stage. A select few are grown at home in the yards and gardens of devotees, though their appearance is fleeting with the autumn frost. Nurtured through the short growing season in this part of Canada, the blooms are an anticipated summer gift, brought to the temple as offerings. Here, regardless of source or season, all are purified with city tap water. This substance too flows into the temple, filtered elsewhere of other agents and impurities. Such is the nature of transnational ritual ecologies cultivated within intemperate climates and clines.

It is tempting to follow Carney and Rosomoff (2009) and characterize this place as another "botanical garden" of the dispossessed. In the next section I consider this possibility, tending to some of the plants and practices indentured laborers brought to British Guiana from India, before turning to those sprouting up in Brampton's suburbs and industrial parks. But to characterize these entities solely by origin or source is far too restrictive, too contained under glass, to account for the many connections, relations, and troubles stirred up along the way (Lowe 2011). By keeping bodies on the move—human, nonhuman, and otherwise—we might better comprehend the transformative potential of touch. As Nazarea reminds us, "Sensuous recollection in marginal niches and sovereign spaces that people carve out of uniformity and predictability constantly replenishes what modernity drains" (2006, 320; Gagnon 2021).

Mother work, in this sense, is embodied work, craft rather than mere commodity. Sensing subtle shifts in tempo and timing, adjusting to obstacles and unstable terrain, requires being light on one's feet. There is, of course, always the risk of getting hurt, and one can count on getting dirty. "You have to go low before you go high," the head priest asserts, assuring me his third eye is wide open.[2] But to go low, to feel grounded, is a hard-fought battle enacted during puja each week. In this space the sovereign meets the sensuous. The sweat and snot, soot and smoke, pounding rhythms and dancing feet firmly emplace rather than merely transcend. To make claims and occupy involves enacting boundaries on *their* terms—fully aware of their position within the state and on the grid. This is plot *life*, in McKittrick's sense, "a creative geography of alternative history practices that has no interest in rehearsing lifelessness" (2013, 11–12).

Dirt, Soil, Earth

> Māriamman is said to be born of earth. . . . Earth is the strongest and humblest of elements; it bears everything, accepting excrement, yielding fruit. People also say that Māriamman has a form of wind, that is, she has no solid body of her own and must take the body given by people, either by possessing them or by entering the stone or earthen images that they make for her. Wind is the embodiment of motion and restlessness and is associated in popular thought with free-wandering spirits and unsatisfied demons, and contrasted with the peaceful higher deities. Wind is also the form of breath and the soul. It is invisible, and so its color is black.
>
> —MARGARET TRAWICK

In Margaret Trawick's study of Tamil Dalit women, she notes that "one of the most important attributes of Māriamman in India is her possession of many forms and many names" (2017, 41). This multiplicity suggests a long history of movement and encounter; hers are affordances that travel well. These conditions, however, make it difficult to stabilize an entity in word and substance, time and place. And yet for millennia, many have tried to do so. Such tenacity might offer a humble lesson for a world steeped in crisis.

Equally amorphous and omnipresent, our current era has been called by many names: Anthropocene, Capitalocene, Chthulucene, Plantationocene, among others (Chwałczyk 2020; Haraway 2016). It is difficult to get one's head and hands around such a beast. As Donna Haraway notes, we may need many names to do so. Within anthropology, Tsing and her colleagues have been prolific contributors to this debate. Their recent proposition is a model they call the "'patchy Anthropocene,' that is, the uneven conditions of more-than-human livability in landscapes increasingly dominated by industrial forms" (Tsing, Mathews, and Bubandt 2019, 186; see also Chao 2022). To grapple with multiple scales and nebulous forces across time and space, they encourage us to pay close attention to the ground. Theirs is a call for an anthropology that "takes landscapes as its starting point and that attunes itself to the structural synchronicities between ecology, capital, and the human and more-than-human histories through which uneven landscapes are made and remade" (2019, 186).

To attend to this unevenness, they highlight two anthropogenic forms: modular simplifications and feral proliferations. Plantations personify this model. These landscapes are dominated by monocrops that thrive on displacement, discipline, and unfree labor. Within the margins, however, diversity gathers in unruly assemblages not so easily contained (Tsing 2012, 2015). Like Wynter's plantations and plots, these forms are not discrete but relational entanglements that shape each other in often unpredictable ways. The approach taken by Tsing and her colleagues differs, however, by emphasizing that "this regimentation of human and nonhuman life must be thought together" (2019, 189). By decentering "the human," other entities can participate more fully in historical processes and be recognized for the contingencies they stirred up along the way. Sugarcane, for example, should be considered "a key participant" in the dynamics of "European conquest and development" (Tsing 2012, 148). But, as Tsing illustrates, before sugar accelerated the plantation economy, rum smoothed the way. This spirit emerged in the seventeenth century on sugar plantations when local yeast spores "contaminated" cane juice and molasses from milling waste (Mintz 1985; Tsing 2012). In this way we can see how "ecological simplifications invite 'feral proliferations' that end up rippling through the entire landscape mosaic" (Tsing, Mathews, and Bubandt 2019, 189).

To see how far these ripples might take us, I turn to one of the lesser-known plantation mosaics, on the northeast coast of South America in what is now Guyana. By 1616, Dutch colonists began occupying the coastal lowlands and importing enslaved West Africans to sculpt an unruly landscape (Hoonhout 2020). An elaborate sea wall and canal system was built on the coast to protect and connect narrow plots of land groomed for cultivation (Vaughn 2022). A few settlers ventured into the interior, a dense jungle that beckoned runaway slaves and harbored Indigenous communities (S. N. Jackson 2012). For a short time, the French would seize control of this land, followed by the British in 1815, who accelerated sugarcane production. Anticipating the abolition of slavery in 1833, British planters solicited the colonial government in India to help "recruit" indentured laborers. The first shipload of workers arrived in 1838, and by 1917 when the indenture system ended, some 239,000 South Asian emigrants had made the arduous journey to British Guiana (Younger 2009, 57; see also Bahadur 2014; Roopnarine 2018).

A midcentury map captures some of these colonial traces in settlement names, property boundaries, and networks of roads and ports (figure 2.2). The narrow strip plantations seem especially artificial when imposed on a sensuous landscape of meandering coastlines and rivers that have little regard for property lines. Meandering, however, rarely yields profits and works at tempos unfit for predictable gains. These things were pushed to the margins and managed in different ways. Following emancipation, enslaved Africans were moved from plantation barracks to nearby settlements, where Protestant missionaries established schools and churches. When indentured laborers arrived, they were settled in the empty slave quarters, or stable-like "logies." These single-story structures were elevated on wood stumps and partitioned into semiprivate rooms that held a family or group of men (Jay-

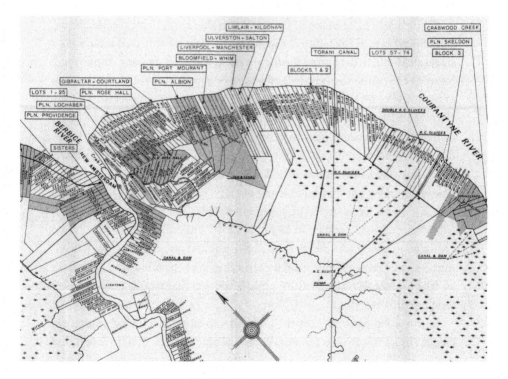

FIGURE 2.2 Map of coastal British Guiana, 1948 (Great Britain, Directorate of Colonial Surveys, *Plan of part of the sea coast of British Guiana showing sugar estates under present cultivation*).

awardena 1963). As with the inhabitants before them, little could be done to alter these structures, though small acts of agency were nurtured outside. Some would prove more fruitful than others.

Indentured peoples brought ganja (*Cannabis sativa*) with them, which grew wild across the Indian subcontinent and was cultivated for various purposes: "making of rope, as medicine, as a dietary supplement, and for religious purposes (e.g., as an ingredient in incense) as well as for its psychoactive properties" (Angrosino 2003, 105). The plant took well to the colony's climate, and for a time, a small cottage industry developed at the plantation's edge. This market was undermined by missionaries, who campaigned against the immorality of its use, alongside the planter bloc, who lobbied for taxes and fees that raised ganja's price, making rum a more attractive substitute (Angrosino 2003, 107). Rum was made cheap and easily accessible in company stores on the estates, which, in turn, created a new cycle of debt and suffering that extended far beyond indenture contracts.

While a salient example of the unpredictable ways "feral proliferations" can continue to reverberate throughout the system, this characterization seems inadequate to capture the full extent of the damage done. Somewhere between the contaminating drift of yeast spores and the managerial position of commodity flows, the ripples developed sharp edges. As scholars of the Caribbean have painfully demonstrated, multigenerational cycles of heavy drinking, poverty, and violence in the region have left deep scars (e.g., Bahadur 2014; Thomas 2011; Williams 1991). Small wonder devotees beg Mariamma for relief from these afflictions—even while offering the divinities who accompany her a hospitable nip of rum. While this ritual act might be portrayed as a lively, more-than-human assemblage inflected with a fungal infection, such a depiction alone elides the lingering violence plantation ecologies engender. As recent critics have pointed out, multispecies flattening obscures the colonial-racial legacies that predetermine who can count as fully human (Davis et al. 2019). In the wake of plantations, many are left longing for this status.

In this light, to be seen by the goddess—to share in a loving mutual exchange, or *darshan*—is a profoundly radical act.[3] It is one of self-recognition and embodied emplacement within a system that treats people as fungible objects (Lowe 2015, 25). Humanness, therefore, is constituted and lived within multispecies assemblages, but in ways that demand recognition of the "fullness of human ontologies" as they are simultaneously articulated and negotiated with alternative forms of life (McKittrick 2006, 122). "This simultaneity," McKittrick continues, "suggests that human geographies are unresolved and are being conceptualized beyond their present classificatory order." To see how people manage to fit into the world and make places for others in it, let us return to plot life in British Guiana.

Unlike the "vile weed," as missionaries characterized ganja, other plants from the Indian subcontinent were welcomed in the colony. On Albion Plantation in 1923, overseer Leslie H. C. Phillips details the more desirable varieties that indentured laborers established there (1960, 41):

> The cocoanut pal (*Cocos nucifer*) and the ubiquitous mango (*Magnifer indica*) have easily won popularity because of their edible fruit; and the beauty of the oleander (*Nerium oleander*) is its passport to many a colonial garden. I suspect, however, that the less-well-distributed nim, or margrosa (*Melia azadirachta*) and pipal, or holy fig (*Ficus religiosa*) were imported for purely religious purposes, as they were then to be seen only around the temples and immigrants' quarters in a few sugar estates.

His account is reminiscent of romantic naturalists a century earlier, in which Latin taxonomies mix freely with religious nomenclature; sustenance and aesthetics are given equal weight; and colonial gardens mingle between places sacred and profane. A lively more-than-human assemblage on paper, but greater than any words or taxonomy might contain. Phillips, nonetheless, tries to make sense of the diversity he encounters within a conceptual framework of the world he inherited. To *know* involved dividing plants up by kind, their source and distribution, as well as use value; the market had yet to gobble these things up. While those who cultivated these plants may not have disagreed with Phillips's account, to know and name these entities, as well as share in worlds with them, involved a different ontology of social relations.

Among the "purely religious" items, neem, or "nim" as Phillips calls it, is an evergreen tree native to the Indian subcontinent.[4] The seeds of the tree would have been easily cached for the ocean crossing. This biogeography of indenture persists today on islands and continents where these trees continue to grow (Chowdhary and Singh 2008, 23). Like ganja, neem is valued for various purposes, but it is widely known for the health and medicinal properties of the bitter leaves. Not unrelated, the tree "is associated exclusively with the worship of the Hindu mother-goddess in her forms as Durga and Mariemmen," Karna Singh stresses in his survey of Guyana's sacred Hindu groves (1980, 25).[5] For some this goes beyond mere association; it *is* the goddess "made tangible in the human world" (Arumugam 2020).

While mother-goddess worship is practiced across the Indian subcontinent, as Mariamma, she is distinctly southern. The so-called Madrasi, who vitalized her worship in British Guiana, were a minority among the vast number of laborers who came predominantly from India's northern central plains. "Madras Coolies" were widely maligned as "indolent, vagrant, and dirty in their habits," as one colonial historian put it (Dalton in Nath 1970). Paul Younger suggests that "in the face of this demeaning criticism—or perhaps partly *because* of it—the South Indians were able to develop a vibrant religious tradition, which is still thriving" (2009, 62). This process, however, was by no means "pure" or exclusive, but pragmatic and opportunistic in its development (Bassier 1978). Once settled in larger numbers, especially in the eastern region of Berbice, "Madras" temples began to proliferate (Younger 2010, 59). In the process, Mariamma's Tamil name became Kaliamma, a more familiar entity to the Hindi majority and inviting of their participation (Stephanides and Singh 2000, 12).

By 1923, this synthesis is reported in Phillips's recollection of the Kalimai puja on Albion Plantation. "In the brilliant light of the full moon," he describes walking from the overseer's compound to a small temple at the front of the estate (Phillips 1960, 40). The path to the entrance was marked by a series of arches made of bamboo poles, each topped with red flags and joined by garlands of "nim, pipal, and mango leaves." The officiating priest, who worked on the plantation during the day, unveiled an adorned statue of the goddess for Phillips to see: "Her face was ebony black and her protuberant eyes—probably two seeds of the mucca-mucca plant (*Caladium arborescens*)—stared fixedly ahead as she sat with her four arms extended laterally holding four objects:—a trident; a sword; the effigy of a human head by the hair; and a hyperboliform native drum" (1960, 40).

Not easily classified as Indigenous or colonial, human, nonhuman, or something else altogether, the mucca-mucca seeds native to South America would, nonetheless, beckon an exchange. Multitudes are found in her gaze. Such abundance was also woven into what Phillips called two "handsome pyramids"—things he was at a loss to name. Instead, he describes their construction with the many imported plants and flowers that had come to grow freely around the immigrant quarters. When finished, the "pyramids" of woven oleander and neem sat atop a small brass pot, or *lota*, on which a small face was molded.[6] Though Phillips did not recognize her, this too was the goddess Mariamma, but in her southern karagam form.

The three-day event would focus on these entities, though they did not stand on their own. They were placed on the heads of two boys, whose bodies she used to dance. This required the "rhythm of the Madras drums," or *tappu*, which Phillips tries to capture by drawing musical scales and notes.[7] But words are all he had to describe the men and women "dancing," the officiating priest "trembling," and the long line of people "waiting" for her help. Like the thundering of the drums, the anticipation of the "wait" is hard to capture on paper. So too are the many pains and sufferings they asked the goddess to alleviate. Phillips informs us that "requests for improved health were most dominant, with relief for poverty a close runner-up. Barren women sought fertility; ambitious young men appealed for more remunerative employment; tormented people begged to be freed from evil," the latter requiring a "stout cane" to beat the "devils" out (1960, 45–46).

The Kali-mai puja culminated by immersing the two karagams in water, setting its pieces adrift. Tendrils of blossoms would animate the sluggish black waters as the current carried them away. But from the destruction, new life might take root in the muddy banks of a canal dug by African slaves. A patch of demonic ground, sublime.

Albion's Seed

Speeding along King's Highway 407 on the western outskirts of Toronto, drivers might easily overlook the gray industrial landscape near the shore of Lake Ontario. For many, this no place is meant to be passed by as quickly as possible to get someplace else. The perception that space is empty or uninhabitable is an extension of plantation logic (McKittrick 2013; Tsing 2015). Not only does this logic elide histories of violence against Indigenous peoples and theft of their lands, but it glorifies developments that make idle space productive. While these hubs may come to teem with activity, they are no place to dwell. "Life, then, is extracted from particular regions, transforming some places into inhuman rather than human geographies" (McKittrick 2013, 7). How must it feel to occupy places deemed lifeless, pushed to the charnel grounds, as it were, to wander with the dying and the dead? "If the plantation, at least in part, ushered in how and where we live now, and thus contributes to the racial contours of uneven geographies," McKittrick asks us, "how might we give it a different future?" (2013, 4). Thomas encourages us to walk alongside, to witness the many struggles and strategies involved in "creating a sensibility of community that disturbs both the spatial and the temporal dimensions of the nation-state through the particular forms of mediation that bring them into being" (2019, 6). Rather than measuring up to perfectionist parameters and ideologies, the "messiness of sovereignty," as Thomas calls it, is experienced in the "everyday ways people innovate life without constantly projecting today's struggle into a future redemption" (7).

By slowing down and meandering along Brampton's side streets and alleys, the innovative sights, sounds, and smells of plot life might be shared. Between the ubiquitous Tim Hortons (Canada's national coffee and donut chain) are "Puja huts," curry houses, sari shops, and vast reception halls, all of which cater to traditions and ritual practices of a vibrant immigrant community (Zhuang 2017, 2019). With over 650,000 people, 81 percent of Brampton's population is a "visible minority," and 65 percent of these are of South Asian descent (Statistics Canada 2021). This census category, however, obscures the many paths taken to and from this place. While most can be traced to colonial histories and postcolonial fractures, the routes are circuitous and diverse (e.g., Ashutosh 2015; Plaza 2004; Premdas 2004; Walton-Roberts 2013).

Since the summer of 2016, I have focused on one of these side streets, a place most would consider on the margins of polite society. The Shri Maha Kali Ammaa Mandir was established in 2001 by a young Canadian of Indo-Guyanese descent. This was the first center of Mariamma worship in Canada, although the association was established using her more familiar Sanskrit name: Kali. The nuances of this history are irrelevant to most who are simply there to see Mother, or "Mudda," depending on how far removed they are from Guyana. The temple, or *mandir*, primarily serves a diverse congregation of first- and second-generation immigrants from the Caribbean and South America. I use the term *congregation* here following Younger (2010), who denotes an unusual style of worship among Indo-Guyanese Hindus in the Caribbean. Not only is their worship characterized by solidarity formed around "kinship ties that give the religious practices a warmth seldom found in any other religious tradition," but, he argues, caste has been replaced "by a very strong sense of the sacred nature of the congregation, something that has little precedent in Hindu tradition" (2010, 91; see also Vertovec 1992). A groupish egalitarian ethos likewise sets the tone at the mandir in Brampton.

Currently, there is a stable core group of about thirty adult devotees; although many others pass through to pray and pay their respects, most are there to seek help. "We are Hindus," the head *pujari* notes, "but we are a healing cult." Theirs is the work of spiritual health—troubles that others will not touch. Pujari, or Jeri, as devotees call him to differentiate his leadership and guru status, explains that the first thing he asks is whether they have been to see a doctor. "If they can't find or treat the problem, it is likely something else, something spiritual that we need to take care of." He is referring to the more-than-human entities and forces that might be carried over from generations past, or more recent evil "put on them" when someone "threw it" their way. Most afflictions manifest through the body—lesions, pains, depression, and menstrual flows—but extend to other troubles as well: marriage and family; immigration status and work; money, debt, and drink.

As a result, there is a transient flow of people through the *mandir*. Some leave revitalized, others dissatisfied, and still others opportunistic, taking what they have learned and opening their own place of worship. It is a schismatic tradition, and followers are known to "shop around." When I asked a long-term devotee about this, why someone would want to leave and take on such responsibility, she just shrugs: "It helps pays the mortgage." While skepticism and suspicion run deep, there is also a pragmatic attitude about the fleeting nature of relations here. Much hangs by a thread, including whether the congregation will continue to have a place to meet.

For a decade now, since 2012, this *mandir* has been in an industrial complex just off a main thoroughfare. During the week, the streets are jammed with semitrucks and the occasional passing train; the area is a snarled mess of tempers and traffic. Reaching the *mandir* on Sunday morning is, by comparison, a breeze. Even then, it is not easily accessed but requires taking a car or bus. The "world perceived through feet," in Ingold's (2011) sense, will have to wait until sandals (or snow boots, depending on the season) are kicked off before going inside. Until then, the mediators between flesh and ground are many. When I first started visiting the *mandir*, a *pujari* made a point of telling me that they serve primarily the poor: "Lots of people don't drive so they may have to take two or three buses to get here. They may not have the money to get home, so we give them $20 to help them buy the return fare; the temple helps them." While this is undoubtedly the case for some, an emerging middling contingent has benefited (however precariously) from Toronto's economic boom.

Women in the congregation are employed in diverse public and private-sector jobs; many are service- or care-oriented positions in health and medicine. A number have pursued postsecondary education and have moved into business and finance. Most men at the temple work in manufacturing, construction, or skilled trades, and a few have started small businesses. Those who purchased homes after immigrating in the late seventies or early eighties have seen their property values skyrocket. Young couples trying to buy a home take on exorbitant debt but hold out hope that the risk will pay off. Many work multiple jobs, and some have taken in renters to help subsidize income. Added to these burdens is the time, labor, and financial demand of keeping a roof over the heads of seventeen deities. Mother's entourage has expanded exponentially over the past century.

To provide a home and attend to their needs requires negotiating shared space within a beige two-story building complex. At the far back of the U-shaped center, a mounted white sign with the *mandir*'s name in red letters marks the front entrance. From the outside, little differentiates it from the other storefronts. Though the faded and curled "OPEN" sign is always inviting, when someone is inside, the heavy glass door is set ajar with a rock from the landscaping—a strip of low-maintenance gravel. Rather than aesthetics, the design is purely functional, meant to absorb runoff and snowmelt. Much of the time, it serves as a catchment zone for cigarette butts, wrappers from the nearby roti shop, or other refuse that has drifted back from the parking lot. Incense smoke curls out through the gap as if trying to clean up the place.

To reach the door requires traversing a sea of asphalt. The parking lot in the center of the complex is carved up into faded white grids, each marked with a number or "visitor." Not unlike Guyana's plantation boundaries, these spaces are designed for efficiency and transience—this is no place to dwell. During the week, the lot serves surrounding businesses—those manufacturing blinds and metal works; redistributing imported foods and goods; and processing laundry on an industrial scale. On Sunday, however, it is given over to people going to "church." This timing, like the congregational shift, is not typical of Hindu worship, but was developed to accommodate work in British Guiana's sugarcane fields and the Christian practices of landowners (Younger 2010, 91). Not unrelated, the parking lot in Brampton is virtually empty on weekends, making it easier to evaluate who, and how, one has "arrived." While blessings are on display, suspicions are also raised. The BMW, or Beemer, is the "dream car" of young men, and driving them at breakneck speeds is part of the local culture of road performance. When Jeri purchased one of these cars a few years back, the gossip became so rampant that he had to get rid of it. "I can't drive that," he concedes. "People think I used their money."

Keeping tabs on one another is a common source of friction in small religious institutions such as this (Dempsey 2006; Vertovec 1992). Maintaining symmetrical relations takes work, particularly in an economy that rewards hierarchy, differentiation, and status. Even with its gifts and utopian promise, an egalitarian ethos requires boundaries and disciplinary regimes; violence takes different shapes. Internal rifts develop and often cut deep, but the greater threat to the collective has more recently come from outside. Competing sects have used "dark work" to inflict harm, and property management has ramped up harassment; these are not necessarily perceived as different forces. Parking has been a chronic source of friction, and it is common for Sunday service to begin with Jeri telling people to go move their cars to a visitor's spot. "You can't just walk a few more feet? I think you'd drive your cars around the different stations if you could," he chides them, waving his arm around the circuit people take to pray and make their offerings each week. Their stay here is becoming increasingly untenable, and they are just waiting for Mother to give the word. It will not be the first time she has done so.

Bhakti in the Burbs

Nearly two decades ago, the first temple was built in the basement of Jeri's family home, a short distance northwest of its current location. Here, he grew up with his brother in a neighborhood surrounded by cornfields in the country. The area has since been engulfed by suburban sprawl (Keil and Üçoğlu 2021). One monocrop has given way to another. Subdivision after subdivision consumes farmland ever farther west, uninhibited by open space. While developers have attempted to give each enclave a different feel, these are just minor variations on a theme: small lots with large homogeneous homes wedged in to maximize all available space. Typically, a small playground sits at the center of each development, led to by meandering sidewalks. An adjacent strip mall offers basic needs—gas, takeout, clinics, and immigration lawyers. While not exactly the "plantation" system Tsing

initially laid out, in many ways it fits the bill: "Plantations are ordered cropping systems worked by non-owners and arranged for expansion. Borrowing
from state-endorsed cereal agriculture, they invest everything in the super-
abundance of a single crop. But one ingredient is missing," she adds. "They
remove the love" (2012, 148).

In the past decade, the average price of a house in Brampton tripled to
over 900,000 Canadian dollars, with 80 percent of buyers carrying a mortgage. More caretakers than owners, the occupants live in a precarious state.
Repossessed properties are quickly snapped up, tidied up, and flipped for a
profit. Others try to hold on by filling basements and beds with renters and
Airbnb guests—myself included. A prepandemic headline from a story in
the *Globe and Mail* encapsulates the situation: "How Canada's Suburban
Dream Became a Debt-Filled Nightmare" (Younglai and Wang 2019). The
most telling statistic in the story's detailed analysis is the debt-to-income
ratio. "In Flower City, half of the labor force is employed in manufacturing,
wholesale trade, retail, transportation, food services and accommodation,
and the median employment income was $31,399 in 2015, nearly $3,500 less
than in Toronto" (Younglai and Wang 2019). Wherein lies the love, indeed.

And yet a planation economy, as we have seen, is never thorough or complete. "In urban jungles as well as rural backwaters, the jumble of diversity
that imperial planners tend to consider excessive still teems" (Tsing 2012,
151). A walk around any of these housing developments is bound to reveal
multicolor triangular flags mounted on bamboo poles, embedded in the narrow strip of land at the front or side of the house. This is a sure sign that an
Indo-Caribbean family lives there. *Jhandis*, or victory flags, are raised in recognition of the divine's power following a thanksgiving puja (Younger 2010,
79, 92). "Planting" these flags not only recognizes a successful communion,
but sacralizes the soil in which they are placed.

Other less obvious aspects of landscaping rituals include the bevy of flowers and plants devotees cultivate as temple offerings. Immigrant homegardens
are, of course, multivalent in purpose. They are a source of well-being and
relief from daily stress, a place of memory and play, along with intergenerational cultural transmission (Mazumdar and Mazumdar 2012). Gardens are a
source of identity and pride but potentially a source of friction; neighbors may
have quite different tastes. The front yard is often structured for these wider
publics to view, while the back becomes a space of more liberatory practice.
The transition to and through these domains offers a sense of agency and

the potential for world making, especially in places where a body feels out of sorts. Cultivating plants can serve as an anchor, however shallow, particularly when land claims are ephemeral or in doubt (Strange 2019). Whether planted around logies, on windowsills in pots, or in the yards of single-family homes, to turn up dirt is an act of emplacement. It is an opportunity to choose with whom to live—however slightly—and how to dwell, however shortly. For devotees in Brampton, careful planning and tending of "temple flowers" during the short growing season is a concerted act of devotion, or *bhakti*.[8] The process can be onerous, requiring time, expense, and effort beyond the long days of work and commute. But, as a devotee reminds me, "if you love, no work."

Negotiations involved in plot living and loving can be seen in my visit with a young family from the congregation who had just moved into their new home. The house is in an older suburb and in need of work; the plan is to fix it up, flip it, and buy something newer and bigger to the west. Warmly greeted by the entire family at the front door, rather than a tour inside, we went straight away to the backyard. They had just finished clearing out the overgrowth and were anxious to show me what had been salvaged from the weeds. Nana, R's mother, insisted that I see the vegetable garden first. Here she had exposed lettuce and onions still thriving, alongside berry vines weaving their way through the chain link fence. R led me on a tour of the ornamental plants, touching them gently, naming each, some by their common names, others by more formal nomenclature. "I'll plant some more things, but for now they just need a little attention." Rather than an act of ownership, she initiated a new relationship based on mutualism and care. The plants will return the gift.

Soaking up the warmth from the concrete and bricks against the house, the flowering shrubs (*Hibiscus syriacus*) were already accommodating. This plant originated in east Asia but spread across the continents as a result of its highly adaptable nature. In some places it is considered invasive because of the ease with which it proliferates (Royal Botanic Gardens, Kew, n.d.), but in Brampton, this vigor is welcomed. For Hindus, the sacred attributes of trees and plants are important, including varieties of hibiscus, which hold special meaning. In a survey of immigrants resettling in Southern California, Mazdumar and Mazdumar describe how "one family's decision to buy a home was partially influenced by the profusion of red hibiscus flowers (considered auspicious) existing in the backyard" (2012, 260). This is particularly the case for Kali worshippers, Goody (1993) notes, in his wide-ranging *The*

Culture of Flowers, but this flower "is not to be found in any Krishna temple for it has a tongue which Kali put out when she stepped on Shiva" (1993, 342). One young woman at the *mandir* has a large tattoo of a hibiscus on her arm and regularly posts images of these dramatic blooms on Facebook. Though she continues to live with her parents, she is beginning to cultivate her own plot living in flesh and digital space.

But R's young children, settling into their new home and backyard, were just getting to know the lay of the land. Her daughter, a beautiful and precocious six-year-old, was the first to point out the purple and pink blossoms on the shrubs. The next day she brought a basket of them to the *mandir*, letting me know that she had helped her mother pick them that morning. Her little brother, however, unimpressed by the attention the plants and his sister were getting, grabbed my hand and, with a full-body tug, "aunty, aunty, come," pulled me toward the patio. As he pointed and babbled something I didn't understand, R interpreted, "Oh, he wants you to see the new barbecue." These things may appear to be mere bourgeoise trappings, signs of manipulative markets and competitive class forces exerting control. At the same time, these "savvy collaborators," as John Burdick calls working-class landscapers, take "advantage of the forces of gentrification to better realize their own life projects" (2020, 449). And while such projects may appear to map onto those of their neighbors—the smell of meat on the barbecue and the aesthetic of a well-groomed yard—the senses might be confused. Collaborations with plants, animals, and more-than-human entities in these plots animate worlds and prospects others might find unimaginable (Dove 2019).

As this young family contemplated making and growing a future, in the more mature neighborhood where Jeri's parents first purchased their home, memories and equity ran deeper. Sitting on their back patio on a Saturday afternoon, Mrs. M and I chatted while she cooked on an outdoor gas stove that her husband, Pops, had built for her. "I do all my temple cooking out here," she told me, while preparing food to sell in the *mandir*'s canteen the next morning. Various dishes rotated through the menu, speaking to the mingling of tastes that emerged in colonial Guiana: Spanish rice, chow mein, black-eyes, chana, and bara (a savory fried fritter). Black-eyed peas were just one of the many plants enslaved Africans brought to the Caribbean and cultivated in their gardens (Carney and Rosomoff 2009, 125). These legumes were well established by the time indentured laborers came to be planted alongside them, but they were welcomed nonetheless, as strange familiars in new homelands.

Surrounding us was a beautifully groomed yard—also Pop's handiwork—with its freshly clipped lawn, mature trees, and flower beds. All harbored stories. Their *jhandi* flags were planted firmly in the ground, with a decorative circle of bricks around the base. Sitting on it were clear plastic cups filled with dirt; one had tipped over with the breeze. "What are you growing?" I asked. "Oh those?" she replied with a chin point while flipping bara in the bubbling grease. "Jeri is trying to grow neem." I went over to set the tipped one upright and to get a closer look at the shoots. At this point, I realized that I had never seen neem that was *growing*. Rather, I was used to the produce guy dropping off big boxes of cut stems at the *mandir* on Saturday afternoons. The cardboard containers were not from Guyana but the "product of" some other tropical clime (figure 2.3). Filled with cut branches, the contents are dumped onto a cloth on the temple floor to be cleaned and sorted into bundles for puja. These looked nothing like the fragile green sprouts peeking through the soil in the cups. The northern climate, of course, is not conducive to the kinds of plants devotees would

FIGURE 2.3 Devotees cleaning and processing imported boxes of neem branches for puja. Photo courtesy of Lars Rodseth.

prefer to grow, but this does not prohibit attempts at "hopeful disappoint-ments" (McKittrick 2013, 12).

The disappointments have been many over the years, as Mrs. M revealed when our conversation slipped into temple troubles. "I don't know what we're going to do," she said, "but Mother will provide." Their moves over the past two decades are rehearsed to confirm this truth. After two years in the basement, the neighbors started to complain about parking as the congregation grew, so they moved to an industrial unit farther south. This arrangement lasted only a year because the owner of the neighboring unit would bang on the walls during service. They then purchased a little slice of heaven—a Korean Christian church for sale on Albion Road. This sacred space would be the congregation's home for eight years until complaints about noise from the drums forced them into the unit where they are now.[9] "We've been through so much . . . sacrificed so much" she said sadly but, with defiance, was quick to add, "This is Mother's house," and again, outlining the structure with head and eyes, "*This* is Mother's house." The implication being that *this* is the source of capital that has allowed them to hold on while being forced to move along.

Playgrounds in Wastelands

"I wish you could have seen the big pujas there!" a *pujari*'s wife reminisced while we sat on the *mandir* floor cleaning neem for the upcoming service. "So many people came to Albion; they came from all over Canada, New York, Florida, and even some from Guyana. We had to make neem bundles and malas all week because so many of them would manifest." She is referring to the three-day Karagam Puja (or Big Puja) that takes place every summer, which became possible only when they moved to Albion Road. Across the street from the church is a city park with a stream running through it, which allowed them to perform the necessary waterside ritual. "The police would block the road so the procession with Mother could cross from the river to the mandir. . . . She comes from the water, you know! It's not like what you see here, where we have to call up, release her, and then drive to the mandir and call up in the parking lot again."

These comments highlight important gains and losses that have occurred over the years. And while they too are hopeful disappointments, a balance sheet logic will not suffice. Rather than transactional, these interactions leave

traces in memories, bodies, and soils. The effects—the affects—of these entanglements are not easily undone. Longing performs its own kind of work.
With the move to Albion Road, not only were vital elements of the landscape
co-present and accessible by foot, but this assemblage animated and drew
together a scattered diasporic community. This gathering, moreover, was
publicly acknowledged and legitimized through the simple act of disrupting
traffic along a main thoroughfare. Others would have to wait or move around
them for a change. Yet one step forward and another one back seems to be
the rhythm of living and loving on the margins. It is, nonetheless, a tempo
and timing all their own. It is the cadence of footwork used by the marlos
and devotees when they dance.

At the current temple location, however, the dancing and "call up" take
place not in the parking lot that serves the storefronts, but rather on a narrow
strip of pavement that circles around the complex to the back. A line of loading
docks provides access to the rear of each shop; when closed, it is difficult to
tell one business from another. On weekends, however, when the center door
is lifted, there is no doubt that you have arrived someplace else (figure 2.4).

The gaping entrance is bracketed by a large blue dumpster on one side
and a pile of sandals and sneakers on the other, clearly delineating sacred

FIGURE 2.4 Back lot during karagam festival, animated by deities dancing the
marlos and karagam boys. Photo courtesy of Maeve Bassett.

space. Four or five men at any given time sit in white plastic chairs, smoking and chatting. Some are still in street clothes, while others have already changed into their yellow *dotis*—five-yard cloth wrapped into knee-high pants—and their temple T-shirts. The latter are printed with the association's logo, address, phone number, and email address; it is a mobile advertisement, though meant only to be worn while doing temple work. All are yellow—Mother's favorite color—except the "white shirts," or the senior *pujaris* who lead the service. Women also pull on these yellow T-shirts, adding a layer to the brightly colored salwar most wear; a long tunic and "Punjabi" pants are the preferred temple attire, especially by those who *play*.

To play, or dance, is a therapeutic act, which involves becoming overwhelmed by the rhythmic vibrations of a deity and falling into an ecstatic trance. Men and women alike "vibrate" with the shakti that has been released, the essence of Mother that all are believed to carry within. The appropriate place to play, however, is "at the back"; like the backyard patio, this is a transitional zone, not quite in but not entirely out. Here, just inside the dock door, is a concrete floor about three meters deep with a large grated drain in the middle. A few chicken feathers, neem leaves, and flower petals remain stuck around the edges inside. This is where call-up occurs, where Mother and other deities are asked to manifest in the bodies of the marlos. Above is a thatched "roof" made from small bamboo shoots woven through wires. The area is simply referred to as "the back," where "outside" work can take place. This inside-out is multipurpose: it allows puja to be performed in the winter with the loading dock door closed to keep the snow and freezing temperatures out, and it obscures the view of management driving by to keep an eye on what they are doing.

The "entrance" to the *mandir* is marked by a large wooden arch; at each end is a flag raised on bamboo poles. Giant pots filled with dirt hold them upright. These are joined by a *torno*, a garland of colorful strips of fabric alternated with mango leaves and stems of neem. A treacherous hump of concrete delineates the transition to "inside," felt when the feet touch a durable indoor-outdoor carpet. This surface, however, is of little note when coming face-to-face with Mariamma. On the main altar platform in the east, she sits, surrounded by her "sisters." Her face pale, her eyes hazel, she stares fixedly ahead with her arms extended. In her hands, she holds a trident, a sword, a doll's head by the hair, and an hourglass-shaped *udukkai* drum. This drum is considered a passage or corridor through which Mother travels. Pujaris tap and sing into its skin, which emits an ethereal sound that charms and coaxes her along. Importantly, as Jeri pointed out, she does not come alone: both good and evil forces accompany her. Three Masters, or protector deities, are there for such purposes, standing at the entrance and by her side. In return, each is offered life sacrifice, cigarettes, and rum, along with sweets; one is fond of black-eyed peas.

But before such hospitality is shown, others must be hosted, beginning with the earth and the sun: Bhoomie Maa (Dhartimata) and Surajnarayan. "Without *them*," Jeri emphasized, "we'd have none of this to offer," sweeping his arm across the devotees on the ground, who are busy cutting up fruit and stringing long strands of flowers, or *malas*, to prepare as offerings. The *malas* will be hung around the necks of the murtis, and their making is a source of admiration and pride. Colors and scents are selected to suit the deities' liking; for those who heal, neem leaves are included in the design. The aesthetic, in general, is an expression of creative skill and taste. A sense of self into substance is cultivated through craft and care (Seremetakis 1994). The raw materials available to do so, however, are influenced by the season, which for much of the year means relying on carnations. Even the kitschiest of flower can be transformed into a thing of beauty. Mrs. S buys the multi-color stems in bulk from the wholesaler on Saturday and sells the buds on Sunday mornings in small baskets. In Brampton's industrial park, there is no hope of establishing a sacred garden like those that grow around temples in the Caribbean (Prorok and Kimber 1997); even dirt must be imported, so thorough is the blight. Life comes from elsewhere each week, extracted from plots and plantations both near and far. Spectral supply chains of global capitalism are necessary evils—unseen, though a presence that is felt. Hard-earned cash becomes the sacrifice of self and substance.

In the summer months, however, the altars burst with bounties from gardens and yards. One aunty is known for growing flamboyant dahlias, large showy buds that bloom late in the summer. The annual puja she sponsors is scheduled around the timing of this plant. Recently, she has had problems with people stealing these blooms from the front of her home, a residence she shares with several renters. While she was doing yardwork she stepped inside for a moment, and "someone took the bud," she told me, shaking her head. But it was not simply a transgression of boundary and property that troubled her; it was a failure to recognize the real value of what they had taken and the damage done: "The dahlia you have to cut, but they broke the stem!" Her disdain is for the lack of knowledge about how to work *with* this entity, how to properly receive a gift that is rewarded through mutualism and care. Rather, the flower was treated like a product plucked from the shelf, with disregard for life itself, and the many entanglements that entails.

Jeri asked the growing crowd, "Why do we make malas?" "Out of devotion!" one young woman excitedly exclaimed, to which he replied, "Yes . . . but to meditate. When you are making malas you should be meditating, not talking and gossiping." A mix of lesson and chastisement, it was an attempt to rein in what he saw as the source of conflict within the congregation. Men and women were active participants—as was the anthropologist encouraging the chatter—but women were characterized as being the more active purveyors of it. While true, it is unusual to see a devotee stringing a mala who is not chatting with someone else; the congregational tenor of the practice relies on these exchanges to weave and reweave cooperative efforts together, which can involve splitting others apart. In the summer months, tensions and gossip are exacerbated by the accumulating expense of sponsoring personal pujas. During this time, the church and activities on Albion Road are often recalled with nostalgia, especially by those who were born in Canada or entered the tradition there. But those who grew up in Guyana long for a different landscape.

"It was easier to contribute because so much was naturally available," a devotee vigorously complained. "You could pick flowers, coconuts, limes, neem, everything you needed for the ritual!" Aunty R was raised in the tradition and attended a *mandir* that was an offshoot of the original temple on Albion Plantation, located just upstream. "Here, you have to purchase everything," she added. Likewise, a young woman who grew up attending a *mandir* just inland from the Albion Estate had similar reflections. She elaborated on the beauty of the temples, how each murti has "their own little home in the compound, and the flowers are everywhere. Hibiscus and oleander— Mother's flower—grow all over and it's so easy to just pick them for puja."

But beyond flowers and fruit, few reminisced about Guyana as a whole. When I asked those who immigrated whether they missed it, they usually shrugged with indifference. "We don't know anyone there anymore; everybody left." Those who did return to see family complained about how dangerous it had become. One young woman even snarled with contempt, "I hate that place and will never go back," while another told me, "It's too hot. I've adapted to the climate here." Even second-generation devotees seem to have little desire to visit the old, new homeland. "I'm a Canadian; I can't even imagine going there—too many bugs." In general, there is consensus that the government is corrupt, Georgetown is full of criminals, jealousy is rampant, and the natural beauty and abundance of the place goes unappreciated. The aunty who grows dahlias came to Canada as a young woman from Berbice. There, her father was a rice farmer, and she summed up her frustration with the place in quite specific ways: "All the food comes from Blackbush, but they don't take care of the canals. The police take bribes and then change their clothes and rob people at night. The land is so beautiful and has so much, but they destroy it"—with an added tooth kiss to emphasize her disdain.

At about 11:00 a.m., four hours after the *mandir* first opened, the *tappu* drums come off the hooks, and Jeri announces opening prayer: "Let's all close our eyes, clasp our hands, and center our hearts on the Universal Mother." Apparently, there was no need to move cars that day. Finally, then, after all the hard work—all the love—the time had come to play.

When the COVID-19 crisis hit in the spring of 2020, quarantine measures in Canada threatened to disrupt the annual cycle of personal pujas. This rupture, however, created new obstacles and opportunities. When it came time for the Big Puja in early July, the city park on Albion Road was closed and gated, so the necessary waterside work could not be done there. Instead, for the first time, the men built the karagams on property the temple association had purchased back in 2006, located farther north, in the cornfields and countryside. Their intention had always been to build a temple compound there, so "the work could be done right." Raising funds and getting permits had made the process long and arduous, but the proliferating effects of the virus nudged them into developing a small patch of this land.

Along the banks of a small stream that meanders through the southern edge of the property, they cut back weeds and removed the overgrowth so that sod could be laid. Buckets of water were filled and brought up from the creek, a splash of turmeric and flower petals added. The *torno* was hung from the surrounding trees, its strips of colorful cloth fluttering in the night breeze beneath the full moon. The handsome pyramids of imported oleander and neem were placed on the heads of two karagam boys, who danced to the thundering rhythm of the *tappu* drums. Eventually, all were loaded into cars and driven back to the *mandir*, where a small group joyously awaited them in the rear parking lot. Leaving *jhandi* flags planted in the muddy bank of *this* stream was quite the victory.

Those of us trapped on the other side of the border or at home because of risk were kept abreast with footage and messages posted on Facebook. Our eyes and lungs were undoubtedly the better for it, but it did not feel the same; the body longed to be there. In a phone conversation the following week, a senior *pujari* gushed about how it was the best Karagam Puja yet. "It was so nice, so remote, like being in Guyana."

Acknowledgments

My sincere thanks to Terese Gagnon for her kind invitation to contribute to this volume and for her thoughtful comments and guidance along the way. I thank Meredith Ellis, Pamela Geller, and Alanna Warner-Smith for their encouragement and close readings of multiple versions of this manuscript. Maeve Bassett, it was a joy working with you on photo documentation. I am most grateful to Karna Singh for sharing his fieldnotes and obscure manuscripts with me. To the many friends and acquaintances I have made at the Shri Maha Kali Ammaa Mandir and beyond, my deepest gratitude for making time for me, sharing your knowledge and stories, and keeping me nourished over the years in such gracious ways.

Notes

1. The spelling I use for Mariamma follows the Guyana Kali Puja lexicon (Stephanides and Singh 2005). Variation is common in and between temples, as the tradition is an oral one that has been passed down through practice.

2. These officiants are called *pujaris*, lay leaders in the temples that facilitate puja services and conduct healing, or *jari*, for those in need. Both men and women can be *pujaris* in the tradition, though the vast majority are men.

3. Meaning "sight" or "vision," *darshan* is an exchange of gaze "that brings good fortune, well-being, grace, and spiritual merit" (Fuller 1992, 59; see also Eck 1981).

4. The current classification of the tree has been changed to *Azadirachta indica* (Royal Botanic Gardens, Kew, n.d.).

5. In Singh's salient study of temples and mosques, he attributes his understanding of the sacred flora to "Pandit Ganesh Ram, an old and learned Hindu priest in Guyana" (1980, 64).

6. The pots, also an import, represent the primordial womb (Stephanides and Singh 2000, 260). In this section, I rely on the glossary in Stephanides and Singh,

Translating Kali's Feast, which was created by Karna Singh, based on his extensive interviews in the 1970s with older *pujaris* and devotees who have long since passed.

7. A *tappu* is a large, circular, and shallow drum usually covered with goat skin. It is carried on the shoulder with a strap and played with two sticks: one a small flat bamboo reed, the other made of thick wood. It is considered the dominant cult drum of the Kali-mai tradition (Stephanides and Singh 2000, 163).

8. Fuller explains that in the popular imagination and in practice, in the sense that "a loving and personal devotion to god has the highest value . . . *bhakti* is the path that brings a person closest to the divine presence and leads most surely to liberation (*moksha*) from the cycle of rebirth, commonly portrayed in devotionalism as entry into heaven to rest eternally at the deity's feet" (Fuller 2004, 157).

9. See George (2018) for a discussion of sonic materialities and the importance of these drums to diasporic communities in North America.

References

Angrosino, Michael V. 2003. "Rum and Ganja: Indenture, Drug Foods, Labor Motivation and the Evolution of the Modern Sugar Industry in Trinidad." In *Drugs, Labor, and Colonial Expansion*, ed. William Jankowiak and Daniel Bradburd, 101–16. Tucson: University of Arizona Press.

Arumugam, Indira. 2020. "Turmeric and Neem: Sacred Herbs, Disease Goddesses and Grappling with Epidemics in Popular Hinduism." CoronAsur: Religion and COVID-19. Asia Research Institute, National University of Singapore. July 22. https://ari.nus.edu.sg/20331-31/.

Ashutosh, Ishan. 2015. "Replacing the Nation in the Age of Migration: Negotiating South Asian Identities in Toronto." *Fennia* 193:212–26.

Bahadur, Gaiutra. 2014. *Coolie Woman: The Odyssey of Indenture*. Chicago: University of Chicago Press.

Bassier, Dennis. 1978. "Kali-Mai in Guyana: Its Spatial Distribution and Divisions." *Release* 3–4:61–77.

Biardeau, Madeleine. 1989. "Brahmans and Meat-Eating Gods." In *Criminal Gods and Demon Devotees: Essays on the Guardians of Popular Hinduism*, ed. Alf Hiltebeitel, 19–33. Albany: State University of New York Press.

Bloomer, Kristin C. 2018. *Possessed by the Goddess: Hinduism, Roman Catholicism, and Marian Possession in South India*. New York: Oxford University Press.

Burdick, John. 2020. "Waging Class Struggle with Plants: Intro-Class Differentiation and Greening Labor in a Public Housing Project in Rio de Janeiro, Brazil." *City and Society* 32:448–74.

Carney, Judith A., and Richard Nicholas Rosomoff. 2009. *In the Shadow of Slavery: Africa's Botanical Legacy in the Atlantic World*. Oakland: University of California Press.

Carney, Judith A. 2020. "Subsistence in the Plantationocene: Dooryard Gardens, Agrobiodiversity, and the Subaltern Economies of Slavery." *Journal of Peasant Studies* 48 (5): 1075–99. https://doi.org/10.1080/03066150.2020.1725488.

Chao, Sophie. 2022. "(Un)Worlding the Plantationocene: Extraction, Extinction, Emergence." *ETropic: Electronic Journal of Studies in the Tropics* 21 (1): 165–91.

Chowdhary, Anjana, and Vinod Singh. 2008. "Geographical Distribution, Ethnobotany and Indigenous Uses of Neem." In *Neem: A Treatise*, edited by K. K. Singh, Suman Phogat, Alka Tomar, R. S. Dhillon, 20–33. New Delhi: I. K. International.

Chwałczyk, Franciszek. 2020. "Around the Anthropocene in Eighty Names—Considering the Urbanocene Proposition." *Sustainability* 12 (11): 4458. https://doi.org/10.3390/su12114458.

Davis, Janae, Alex A. Moulton, Levi Van Sant, and Brian Williams. 2019. "Anthropocene, Capitalocene, . . . Plantationocene?: A Manifesto for Ecological Justice in an Age of Global Crisis." *Geography Compass* 13:1–15.

Dempsey, Corinne G. 2006. *The Goddess Lives in Upstate New York: Breaking Convention and Making Home at a North American Hindu Temple*. Oxford: Oxford University Press.

Dove, Michael R. 2019. "Plants, Politics, and the Imagination of the Past 500 Years in the Indo-Malay Region." *Current Anthropology* 60 (S20): 309–20.

Eck, Diana L. 1981. *Darśan: Seeing the Divine Image in India*. Chambersburg, Pa.: Anima.

Fellows, Kristen R., and James Delle. 2015. "Marronage and the Dialectics of Spatial Sovereignty in Colonial Jamaica." In *Current Perspectives on the Archaeology of African Slavery in Latin America*, edited by Pedro Funari and Charles E. Orser, 117–32. New York: Springer.

Fuller, C. J. 1992. *The Camphor Flame: Popular Hinduism and Society in India*. Princeton, N.J.: Princeton University Press.

Gagnon, Terese V. 2021. "Affective Ecologies: Karen Plant Companionships Across Borders." In *Moveable Gardens: Itineraries and Sanctuaries of Memory*, edited by Virginia D. Nazarea and Terese V. Gagnon, 19–46. Tucson: University of Arizona Press.

George, Stephanie Lou. 2018. "Invoking the Supernatural and the Supranational: *Tappu*, Trance and Tamil Recordings in Indo-Guyanese 'Madras Religion' and the Politics of Sonic Presence." *Civilisations* 67:41–56.

Goody, Jack. 1993. *The Culture of Flowers*. Cambridge: Cambridge University Press.

Graeber, David. 2011. *Debt: The First 5,000 Years*. Brooklyn, N.Y.: Melville House.

Haraway, Donna. 2015. "Anthropocene, Capitalocene, Plantationocene, Chthulucene: Making Kin." *Environmental Humanities* 6 (1): 159–65. https://doi.org/10.1215/22011919-3615934.

Haraway, Donna J. 2016. *Staying with the Trouble: Making Kind in the Chthulucene*. Durham, N.C.: Duke University Press.

Hartman, Saidiya V. 2002. "The Time of Slavery." *South Atlantic Quarterly* 101 (4): 757–77.

Hirsch, Alison B. 2019. "Ritual Practice and Place Conflict: Negotiating a Contested Landscape Along Jamaica Bay." In *Routledge Companion on Global Heritage Conservation*, edited by Vinayak Bharne and Trudi Sandmeier, 265–78. New York: Routledge.

Hoonhout, Bram. 2020. *Borderless Empire: Dutch Guiana in the Atlantic World, 1750–1800*. Athens: University of Georgia Press.

Ingold, Tim. 2011. *Being Alive: Essays on Movement, Knowledge, and Description*. New York: Routledge.

Jackson, Shona N. 2012. *Creole Indigeneity: Between Myth and Nation in the Caribbean*. Minneapolis: University of Minnesota Press.

Jackson, Stephanie Lou. 2016. "From Stigma to Shakti: The Politics of Indo-Guyanese Women's Trance and the Transformative Potentials of Ecstatic Goddess Worship in New York City." In *Indo-Caribbean Feminist Thought: Genealogies, Theories, Enactments*, edited by Gabrielle Jamela Hosein and Lisa Outar, 301–19. New York: Palgrave Macmillan.

Jackson, Zakiyyah Iman. 2020. *Becoming Human: Matter and Meaning in an Antiblack World*. New York: New York University Press.

Jayawardena, Chandra. 1963. *Conflict and Solidarity in a Guianese Plantation*. London: Athlone.

Keil, Roger, and Murat Üçoğlu. 2021. "Beyond Sprawl? Regulating Growth in Southern Ontario: Spotlight on Brampton." *disP—The Planning Review* 57:100–118.

Khan, Aisha. 2015. "Material and Immaterial Bodies: Diaspora Studies and the Problem of Culture, Identity, and Race." *Small Axe* 19 (3): 29–49.

Kloβ, Sinha T. 2016. *Fabrics of Indianness: The Exchange and Consumption of Clothing in Transnational Guyanese Hindu Communities*. New York: Palgrave.

Le Petitcorps, Colette, Marta Macedo, and Irene Peano, eds. 2023. *Global Plantations in the Modern World: Sovereignties, Ecologies, Afterlives*. Cham: Palgrave Macmillan.

Lowe, Lisa. 2011. *The Intimacies of Four Continents*. Durham, N.C.: Duke University Press.

Mazumdar, Shampa, and Sanjoy Mazumdar. 2012. "Immigrant Home Gardens: Places of Religion, Culture, Ecology, and Family." *Landscape and Urban Planning* 105 (3): 258–65.

Mbembe, Achille. 2019. *Necropolitics*. Durham, N.C.: Duke University Press.

McKittrick, Katherine. 2006. *Demonic Grounds: Black Women and the Cartographies of Struggle*. Minneapolis: University of Minnesota Press.

McKittrick, Katherine. 2013. "Plantation Futures." *Small Axe* 17 (3): 1–15.

McKittrick, Katherine. 2021. *Dear Science and Other Stories*. Durham, N.C.: Duke University Press.

McNeal, Keith E. 2011. *Trance and Modernity in the Southern Caribbean: African and Hindu Popular Religions in Trinidad and Tobago*. Gainesville: University of Florida Press.

Mintz, Sidney W. 1985. *Sweetness and Power: The Place of Sugar in Modern History*. New York: Penguin.

Nath, Dwarka. 1970. *A History of Indians in Guyana*. London: printed by the author.

Nazarea, Virginia D. 2006. "Local Knowledge and Memory in Biodiversity Conservation." *Annual Review of Anthropology* 35:317–35.

O'Hara, Dale. 2007. *Acres of Glass: The Story of the Dale Estate and How Brampton Became "The Flower Town of Canada"*. Toronto, ON: eastendbooks.

Palmié, Stephan. 2023. *Thinking with Ngangas: What Afro-Cuban Ritual Can Tell Us About Scientific Practice and Vice Versa.* Chicago: University of Chicago Press.

Phillips, Leslie H. C. 1960. "Kali-Mai Puja." *Timehri* 39:40–46.

Plaza, Dwaine. 2004. "Disaggregating the Indo- and African-Caribbean Migration and Settlement Experience in Canada." *Canadian Journal of Latin American and Caribbean Studies* 29 (57–58): 241–66.

Prorok, Carolyn V. and C.T. Kimber. 1997. "The Hindu Temple Gardens of Trinidad: Cultural Continuity and Change in a Caribbean Landscape." *Pennsylvania Geographer* 35 (2): 98–135.

Premdas, Ralph. 2004. "Diaspora and its Discontents: A Caribbean Fragment in Toronto in Quest of Cultural Recognition and Political Empowerment." *Ethnic and Racial Studies* 27 (4): 544–64.

Rodney, Walter. 1972. *How Europe Underdeveloped Africa.* London: Bogle-L'Ouverture.

Roopnarine, Lomarsh. 2018. *The Indian Caribbean: Migration and Identity in the Diaspora.* Jackson: University of Mississippi Press.

Royal Botanic Gardens, Kew. n.d. "*Azadirachta indica.*" Plants of the World Online. https://powo.science.kew.org/taxon/urn:lsid:ipni.org:names:1213180-2.

Royal Botanic Gardens, Kew. n.d. "*Hibiscus syriacus* L." Plants of the World Online. https://powo.science.kew.org/taxon/urn:lsid:ipni.org:names:560890-1.

Seremetakis, C. Nadia. 1994. "The Memory of the Senses, Part I: Marks of the Transitory." In *The Senses Still: Perception and Memory as Material Culture in Modernity*, edited by C. Nadia Seremetakis, 1–18. Chicago: University of Chicago Press.

Singh, Karna Bahadur. 1980. *Temples and Mosques: An Illustrated Study of East Indian Places of Worship in Guyana.* Georgetown, Guyana: Release.

Statistics Canada. 2021. "Census Profile, 2021 Census of Population." Statistics Canada. https://www12.statcan.gc.ca/census-recensement/2021/dp-pd/prof/index.cfm?Lang=E.

Stephanides, Stephanos, and Karna Bahadur Singh. 2000. *Translating Kali's Feast: The Goddess in Indo-Caribbean Ritual and Fiction.* Amsterdam: Rodopi.

Strange, Stuart Earle. 2019. "Indigenous Spirits, Pluralist Sovereignty, and the Aporia of Surinamese Hindu Belonging." *Ethnos* 84 (4): 642–59.

Thomas, Deborah A. 2011. *Exceptional Violence: Embodied Citizenship in Transnational Jamaica.* Durham, N.C.: Duke University Press.

Thomas, Deborah A. 2019. *Political Life in the Wake of the Plantation: Sovereignty, Witnessing, Repair.* Durham, N.C.: Duke University Press.

Thomases, Drew, and James D. Reich. 2019. "'Out-of-the-Box' Hinduism: Double Diaspora and the Guyanese Hindus of Queens, NY." *Journal of Hindu Studies* 12 (3): 339–60.

Trawick, Margaret. 2017. *Death, Beauty, Struggle: Untouchable Women Create the World.* Philadelphia: University of Pennsylvania Press.

Trouillot, Michel-Rolph. 2002. "North Atlantic Universals: Analytical Fictions, 1492–1945." *South Atlantic Quarterly* 101 (4): 839–58.

Tsing, Anna. 2012. "Unruly Edges: Mushrooms as Companion Species." *Environmental Humanities* 1:141–54.

Tsing, Anna Lowenhaupt. 2015. *The Mushroom at the End of the World: On the Possibility of Life in Capitalist Ruins*. Princeton, N.J.: Princeton University Press.

Tsing, Anna Lowenhaupt, Andrew S. Mathews, and Nils Bubandt. 2019. "Patchy Anthropocene: Landscape Structure, Multispecies History, and the Retooling of Anthropology." *Current Anthropology* 60 (S20): 186–97.

van der Veer, Peter, and Steven Vertovec. 1991. "Brahmanism Abroad: On Caribbean Hinduism as an Ethnic Religion." *Ethnology* 30 (2): 149–66.

Vaughn, Sarah E. 2022. *Engineering Vulnerability: In Pursuit of Climate Adaptation*. Durham, N.C.: Duke University Press.

Vertovec, Steven. 1992. "Community and Congregation in London Hindu Temples: Divergent Trends." *New Community* 18 (2): 251–64.

Walton-Roberts, Margaret. 2013. "Introduction: South Asian Diasporas in Canada." *South Asian Diaspora* 5:1–5.

Weheliye, Alexander G. 2014. *Habeas Viscus: Racializing Assemblages, Biopolitics, and Black Feminist Theories of the Human*. Durham, N.C.: Duke University Press.

Williams, Brackette F. 1991. *Stains on My Name, War in My Veins: Guyana and the Politics of Cultural Struggle*. Durham, N.C.: Duke University Press.

Woods, Clyde A. 1998. *Development Arrested: The Blues and Plantation Power in the Mississippi Delta*. New York: Verso.

Woods, Clyde A. 2007. "'Sittin' on Top of the World': The Challenges of Blues and Hip Hop Geography." In *Black Geographies and the Politics of Place*, edited by Katherine McKittrick and Clyde A. Woods, 46–81. Toronto, ON: Between the Lines Press.

Wynter, Sylvia. 1971. "Novel and History, Plot and Plantation." *Savacou* 5:95–102.

Wynter, Sylvia, and Katherine McKittrick. 2015. "Unparalleled Catastrophe for Our Species? Or, to Give Humanness a Different Future: Conversations." In *Sylvia Wynter: On Being Human as Praxis*, edited by Katherine McKittrick, 9–89, Durham, N.C.: Duke University Press.

Younger, Paul. 2009. *New Homelands: Hindu Communities in Mauritius, Guyana, Trinidad, South Africa, Fiji, and East Africa*. New York: Oxford University Press.

Younglai, Rachelle, and Chen Wang. 2019. "How Canada's Suburban Dream Became a Debt-Filled Nightmare." *Toronto Globe and Mail*, September 13. https://www.theglobeandmail.com/business/economy/article-how-canadas-suburban-dream-became-a-debt-filled-nightmare/.

Zhuang, Zhixi Cecilia. 2017. "Creating Third Places: Ethnic Retailing and Place-Making in Metropolitan Toronto." In *Urban Transformations: Geographies of Renewal and Creative Change*, edited by Nicholas Wise and Julie Clark, 97–114. New York: Routledge.

Zhuang, Zhixi Cecilia. 2019. "Ethnic Entrepreneurship and Placemaking in Toronto's Ethnic Retail Neighbourhoods." *Tijdschrift voor economische en sociale geografie* 110 (5): 520–37.

CHAPTER THREE

Storying Against Dispossession

Nurturing Memories of Other Worlds

TERESE V. GAGNON

Alienation from homelands and food systems has been characteristic of life for many residents of Mae La refugee camp in Thailand, located close to the Myanmar border. I witnessed these trends during the six months I lived in the camp in 2018, conducting ethnographic research and volunteering as a teacher at a college in the camp. Meanwhile, through my interviews and friendships with Indigenous Karen young people living in Mae La, I learned that many of them frequently reminisced about their home villages and dreamed vividly of returning there.

These young people, many of whom were my students, in their midtwenties, would often visit me in the evenings to chat and drink tea. At these times they spoke about their dreams of one day returning home to their villages on the Karen side of the border (otherwise known as Myanmar). It was a commonly held goal among these young people to contribute to their local communities and to build their futures in their respective home villages.[1] As aid funding dries up, pressures toward precarious migrant labor and expulsion (Sassen 2010, 2014) are real and palpable in the camp. In the face of this, many of my young friends pushed back against the seeming inevitability of their flowing out of the camp to become migrant laborers in Thailand.

Here, I investigate how these students enacted a politics of refusal, as theorized by anthropologist Audra Simpson (2014, 2017), through engagements with food and biodiversity from their homes. Such refusal was manifested in their affect-laden stories of home, as well as through sharing foods from home, sent across borders in defiance of settler state logics (Simpson 2014).

Indeed, both sides of the border were Indigenous Karen territory before the states of Thailand and Myanmar attempted to extend their administrative reach and put up a border there. Like the process described by Audra Simpson in Mohawk territory (bordered by the United States and Canada), the work of settler expansion and its refusal by Karen communities is very much ongoing. In this specific context of ongoing settler colonialism, students further enacted refusal through their detailed imaginings of future returns (see Nazarea 2021) to their homes left behind. This was even as such returns remained necessarily far off or forestalled because of armed conflict and systemic oppression.[2] The stories they told each other highlighted agrarian and land-based dimensions of rural life, filled with more-than-human relationships and a pervasive sense of greater togetherness than life in the camp offered.

Finally, I suggest that these young peoples' identities as border crossers—both literally and metaphorically—afforded them a powerful perspective from which to critique both settler states and humanitarian regimes. In doing so, they refused the implicit inevitability of their flowing out of the camp to become part of the global precariat (Standing 2011) as migrant laborers in the Mae Sot Special Economic Zone (SEZ) and beyond (see Pongsawat 2007). As I illustrate here, some important ways Karen refugee youth in Mae La camp enacted refusal were through affective engagements with food and plants tied to their homes. These counter-acts and counter-narratives are limited in what they can change about the current political reality and conditions of life in the camp. Yet, their sensory and embodied nature makes them portable and durable sites, or rather *seeds*, of sovereignty . . . primed and waiting for future openings.[3]

"We Have to Spend the Life of the Refugee"

As funding for the camp dries up and it remains unsafe to return to homes in Myanmar and Kawthoolei, the Karen homeland, many young people from the camp have become undocumented migrant labors in Thailand or elsewhere as a means of surviving and supporting their families.[4] As Norum, Mostafanezhad, and Sebro (2016) describe, Mae La is positioned as a complex liminal space between Indigenous Karen homelands and spaces of intensified labor exploitation in the Mae Sot SEZ, characterized by the temporality of late capitalism. Additionally, I argue that despite numerous humanitarian initiatives aimed at preparing refugees for return, life in the camp helps transition Karen refugees from the lived reality of the village to that of the SEZ. This is because, despite the terrible conditions that caused refugees to flee their villages, it is in the camp that most first become accustomed to toiling to survive without freedom of movement or rights *outside the context of armed conflict.*

One teacher at the college where I taught, Thara Ta Tha Wi, had been a member of the "orphan" dormitory when he was a student in the camp. He now runs his own student dormitory, located a short walk from the school. During an interview one afternoon in fall 2018, reflecting on the uncertain future of the camp and the challenges facing camp residents, he remarked to me, "In 2019, 2020, something like that, we know that we will all have to go back [to Myanmar]." We sat in turquoise plastic chairs on the stage of the empty school auditorium, where daily morning chapel is held, my iPhone recorder running on the table in front of us. The afternoon light slanted in through the patterned concrete blocks, and a scrawny chicken pecked around the neatly spaced plants at the threshold. Thara Ta Tha Wi continued, "That's what we heard. We will all have to go back. They will stop the refugee kinds of things. Food is becoming decreased—that is *very* true. That has happened already. That was last year [when there was a cut in rations]."

When I asked Thara Ta Tha Wi how people were coping with the decreased amounts of food, he informed me, "So far, people can still manage, but they know the food is decreasing. Maybe some of them, they earn a little bit, you know. They buy rice, so they can still survive. As long as they can survive in the camp, they will stay here. But if they cannot really survive here, I think *they will go* . . . somewhere else, in Burma or in some Thai Karen villages here. So, when people go, this place will remain as a village . . . not as a refugee [camp] anymore." Thara Ta Tha Wi added after a moment, "They cannot force, if the refugee is not willing to go back. If they force, it is some kind of human rights abuse.[5] That's what I heard. If the refugees want to stay—they have to let them stay. But the pressure will come. With the [cuts in] food—they will give them pressure *that is for sure. That* will make a refugee not feel like staying in the refugee camp. Those kind of pressures already started, and the feeling already started. So, some of them already went back. But *a lot* remain."

A teacher in his midthirties, Thara Ta Tha Wi had more stability and options than most. He lived and taught in the camp as a chosen vocation and could return to his Karen village, located almost directly across the border, where it was safe at that time to go because the fighting in that spot had subsided. Also, unlike many in the camp, he did not have to worry about retribution from the Burmese military if he went back to his village because it was in a Karen-controlled area. His mother still lived in the village, where she kept a cow, and his relatives had durian and betelnut orchards there. The products of these orchards were widely known in the camp and prized for what was regarded as their superior taste. Selling betelnut to camp residents provided Thara Ta Tha Wi's relatives with a significant source of income.

Students from the school were sometime able to informally leave the camp to take weekend trips to his village, something not possible for most camp residents. They went to renew their spirits (see Kantor 2021) in the verdant rural setting, to hike into the woods and go durian picking and *eating*—a special treat indeed. This I learned early on in my stay at the camp, when my friend and fellow volunteer teacher Ruth, a Karen woman living in Singapore, mysteriously disappeared for a weekend. She returned looking refreshed and with stories to tell and pictures to share of lush rural scenes, everything slick and emerald from the abundant rains. Apologizing that she had not been able to tell me where she was going before she left, she explained that she had been invited to join the excursion to Thara Ta Tha Wi's home village at the last minute, being told "get in the truck now if you want to come!"

Ruth let me travel vicariously with her and the small group of students through her vivid stories and images, taken on her Galaxy phone with a photographer's eye. She told me about how her legs and back ached after hours of hiking, but how it was "totally worth it," as the durian they picked in the orchard was, she said, "some of the best durian I have had in my life." This was high praise coming from a durian connoisseur and an all-around food enthusiast. With animation she described the serenity of the village, the animals, and the agricultural fields, the forest, and the clear streams running down from the hillsides. Despite being geographically proximate, the village represented a strong contrast to the camp, with its overcrowdedness and stinky brown water that made our skin itch. She showed me a video clip of the students laughing by the edge of one such stream, a small footbridge over it. In the photo they leaned on each other and gleefully showed off their haul of rotund, spiky durian, reveling in how muddy they had gotten along the way. From these images and Ruth's description of the expedition, it was apparent there was much longing among the students for such rural settings, which mirrored the home villages many of them had needed to leave behind when they came to live in the refugee camp.

Yet, the social safety net and connection to village life that Thara Ta Tha Wi had in his home village—which he could share to some degree with the students—is not available to the vast majority of those living in the camp. For most people in Mae La who lack a clear path home, the dire reality and uncertain future of the camp is perceived clearly. It is felt viscerally in the everyday, where there are progressively fewer recesses in which to take shelter from the constricting difficulty of their circumstances. Squeezed by what C. Nadia Seremetakis describes as the "im/possibility" of daily life (2019, 4) under such conditions of contraction, young people in the camp, such as Saw Khay Thu (a student in my sociology class), reflected perceptively on the forces affecting their experience of simultaneous confinement and precarity.

The phenomenon they described to me parallels the empirical reality and theoretical concept of "emptiness" elaborated by anthropologist Dace Dzenovska (2020). Writing from the context of an "emptying" countryside in Latvia, near the Russian border, Dzenovska transports her reader to her quiet fieldsites: whole villages that have been abandoned by the circuits of global capital. Here, tellingly, even promises of future development are no longer made by officials. She describes how residents of these places fear that soon their villages will cease to exist on the map. So, too, some young people in Mae La camp like Saw Khay Thu fear they may lose their connection to their homes and their Karen identity, which lends meaning to their lives. This fear is also due largely to the routes of global capital, which concentrate in some places and create abandonment in other places and which force labor migration and ensuing "emptiness" on a massive scale.

I talked with Saw Khay Thu one evening after his long day of dormitory chores and classes and before an evening drama performance. During our conversation, he related to me with surprising frankness and vulnerability his vast uncertainty about the future. He confessed to me the loneliness he felt living in the camp, far away from his family in a mixed-control border area of Myanmar. He told me somberly how his mother had to proclaim to local Burmese officials that he had "already died" so that his family would not get in serious trouble for him going to the camp. As such, he could not go home.

An enthusiastic student in my sociology class, Saw Khay Thu related to me his incisive critiques of his and other refugees' treatment at the hands of the Thai government, NGOs, and other groups that made decisions about

FIGURE 3.1 College students celebrate their graduation, yet for many of them, life after this milestone is highly uncertain. Photo by author.

their lives for them. In so doing, he exhibited the keen perspective of a social scientist: a disposition I had picked up on from his serious and sometimes impassioned contributions to class discussions. During our conversation he reiterated the overwhelming constriction he felt at never being able to leave the camp, including to work or to farm, and his despair at having no clear path forward for his future. He told me, "We stay here but everything is not okay. . . . Day by day, week by week, years by years, we have to spend our life in the camp. We have to spend the life of a refugee."

As illustrated by my conversation with Saw Kay Thu, the perceptiveness that young people have about life in and beyond the camp is partially due to their perspective as border crossers. Integrating two separate concepts of "border crossing," I mean that these young people travel between different groups and lived realities (Hromadzic 2011, 276; Feuerverger 2001, xiii) and that in the process of doing so, they enact sovereignty by moving themselves, along with food and stories, across settler colonial borders and boundaries on the landscape, as theorized by Audra Simpson (2014).[6] It is from this unique vantage point as border crossers that the these young people find potent ways of rejecting the implied inevitability of their flowing out of the camp to become part of what Guy Standing calls "the global pre-cariat" (2011). Some of the primary ways I witnessed refugee young people rejecting conditions of dispossession and alienation were through affective engagement with food and plants tied to home. For example, they frequently recounted memories of food and agricultural practices in their home villages, such as working on their families' farms and foraging and fishing in the company of friends.

Sensory Politics and Students' Counter-Acts

The counterpoints to the difficult stories I heard from young people like Saw Khay Thu were the subtle but subversive counter-acts and counter-narratives (Arampatzi 2017) I encountered among young people in the camp. Through sensory and embodied engagement with food and plants tied to home, these acts and narratives contested the dispossession, alienation, and forgetting (see Seremetakis 1994, 2019; Sutton et al. 2013; Sutton 2021) occurring in the camp. Students' counter-acts included foraging in the forest around the camp for wild foods, such as various herbs and wild ginger, as well as for the leaves of teak trees, used for roof repair, and bamboo for house repair. This was even while foraging in the forest was strictly prohibited by Thai authorities at the penalty of steep fines, as well as the possibility of extra-judicial punishments ranging from a harsh beating to even death. In some extreme instances, refugees found cutting bamboo have been shot on sight (Human Rights Watch 2012). Other counter-acts included young people cultivating food against the odds of limited space, limited irrigation, and poor soil quality, as in the case of the "boarders' gardens" cultivated by high school students living in communal dormitories. In such gardens (pictured in figure 3.2), students grew a significant portion of their own food despite severe limitations on space and resources. In that space, agricultural knowledge was perpetuated by being shared peer-to-peer.

Furthermore, students overcame the alienation of humanitarian logics that framed food as a biopolitical tool for survival, as described by Trapp (2016), by finding joy through cooking and eating together. This was despite a chronic lack of sufficient quantities of nutritious food. In this context, acts of harvesting wild foods, gardening, cooking, and tasting food all in the company of others within a shared field of experience countered alienation by connecting students with intergenerational memory and the landscapes of

FIGURE 3.2 Students work in the dormitory gardens, or "boarders' gardens," where they grow some of their own food to supplement reduced rations, and where they share agricultural skill and knowledge with each other. Photo by author.

their rural home villages. In these unspectacular moments, the reinforcement of sensory and embodied memories rooted in Karen homelands was filled with quiet potentiality. This was the potentiality to restore linkages to home even within broad conditions of dispossession and forgetting, which can be read as a form of refusal (Simpson 2014). It is true that these counteracts and counter-narratives are limited in what they can change about the current political reality and conditions of life in the camp. Nonetheless, much like Deborah Thomas's writing on the power of Caribbean diasporic dance as sovereignty (2019), the sensory nature of these counter-acts—tied to landscape and the lives of more-than-human beings—makes them embodied moments of communal sovereignty.

Foods from Home

Spending time with young people in the camp, I came to understand that there was ubiquitous nostalgia for the foods and more-than-human relationships embedded in their fraught rural homelands. This longing was coupled with their biting critiques of "unhealthy," "flavorless," "too sweet," and "chemical" foods in the camp. One day, sitting in the school café, taking advantage of the slow-moving electric fan that drew students and teachers there, a young man, Saw Hey Su Law, told me about the pros and cons of life in the camp versus back in his village.

His father, back in Karen State, was an abusive alcoholic. Coming to the camp had finally allowed him freedom from his father's control. On the other hand, he told me, the quality of life was not good in the camp. The food, he explained, was not fresh like back home. It lacked flavor or had a bad taste, and it provided him with little energy. Additionally, the food in the camp was full of chemicals that caused his skin to break out, something that had never happened in his village. There, he said, the food was incredibly fresh and flavorful, and it gave him lots of energy—most of it being grown locally or harvested from the forest. In his village the water and the vegetables smelled "sweet" and "fresh," unlike the water and food in the camp, which already smelled "old" as soon as one got it.

On a different occasion, while interviewing a student at the kitchen table of our teachers' dormitory with a group of other students sitting around taking part, the young man I was speaking with, Saw La Say, asked me if I liked Thai food. "Yes," I replied. He told me that it was his goal to open a food stand in the camp to make money. He planned to specialize in Thai food because it was easy to cook and popular. Yet, he told me, he did not like Thai food but greatly preferred Karen food. He explained that this was because Thai food was far too sweet for him. Karen food, on the other hand, was filled with different kinds of wild herbs and the pungent fermented fish flavor of *nya u ti* (fish paste), anchored by complex notes of bitter, sour, and spicy. These were only two instances of the numerous times young people criticized the camp food in front of me and contrasted it with the fresh, flavorful, and healthy foods from their rural homes.[7]

Both the nostalgia for foods left behind and critiques of food in the camp parallel Mike Anastario's findings in his work with rural Salvadoran migrants to the United States. In his monograph *Parcels* (2019), as well as his contribution in this volume, Anastario investigates the transnational movement of food and seed packages from home to migrants in states such as Colorado. The parcels Anastario tracks are, for the people who receive them, weighed with the memories of agrarian lifeworlds and Indigenous/*campesinx* ways of knowing: the tastes, smells, and textures of their contents interwoven with the rural landscapes of home. Anastario recounts how his interlocutors called U.S. food "disgusting." Writing during the Trump years, he observes, "some members of the Trump administration doubt whether all Salvadoran migrants deserve to be referred to as human, while some Salvadoran migrants doubt the integrity of what U.S. Americans call food" (2019, 1). This resonates with my students' bold critiques of food in the camp—a space in which their humanity is also perpetually called into question.

Taking this critique of food in the camp a step further was a young student's damning characterization of all of Thailand as "cheap." Reflecting on how she missed her old village in Karen State and found relationships shallow in the camp, the student wrote in her Facebook story, "I miss my old school and my old friends. No wonder people only come to Thailand for readymade clothes and cheap things." I read this as an affect-laden rebuke of the dehumanizing treatment she had faced in Thailand and the ways individualized life in the camp caused a breakdown in feeling of togetherness. Life back home had been objectively bad in numerous ways for these young people. Most notably, it was marked by periods of violent conflict as well as continuous oppression, some combination of which had driven all of them to the border. Yet, this did not mean that Thailand was immune to their scalding critical assessments.

My students' critiques, issued through the language of the sensory, also resonate strongly with various other ethnographic accounts from around the world (see Glazebrook 2008; Gold 2009; Gordillo 2004; Kantor 2021; Seremetakis 1994, 2019; and Sutton 2008). Among these are ethnographies of other protracted refugee situations, such as those detailed by Elizabeth Dunn (2017) and Micah Trapp (2016). Dunn, in her monograph, describes how the Georgian refugees she worked with called humanitarian macaroni "not food" (2017, 95). Similarly, Trapp's interlocutors at Buduburam Liberian refugee camp in Ghana creatively derided their rations as "Drip" and "You-Will-Kill-Me beans" (2016, 417). Reflecting on this, Trapp remarks, "In the biopolitical domain of the refugee camp, food is a site of gustatory discipline" (2016, 413). As she observes, and as my young friends in Mae La proved, food is also a fertile terrain for critique and refusal within this humanitarian biopolitical sphere.

In Mae La, students rejected dispossession and alienation through small subversive acts and by telling stories tied to food and plants from their homes. These sensory embodied acts and narratives helped them re-member (as in reassembling a body) more-than-human relationships embedded in and transmitted through the "sensible sphere" (Rancière cited in Trapp 2016, 432). These same young people used the grammar of sensory politics tied to taste to refuse the less-than-human status they were assigned in the camp. In the stories refugee young people told, they recast themselves as members of a generation enacting continuity, connection, and return to Kawthoolei. These narratives refused their dehumanization and the erasure of Karen sovereignty by surfacing the sensory dimensions of everyday life in their homes, such as the embodied joy of "getting fresh air" while fishing and foraging in the company of friends.

Students' acts of refusal, however, were not limited to breaking the rules about foraging, cultivating gardens communally in a context that demanded individualized self-reliance, or sharing narratives laden with sensory politics. Additionally, and impressively, students moved food across the border erected by Myanmar and Thailand, crosscutting Karen territory, to maintain connections to their homes left behind. This became a way for students to maintain relationships with family members and with the unique biodiversity in their "native places," as students called their homes. Like the students themselves, these foods from home crossed the settler colonial border created by the states of Thailand and Myanmar. In so doing, they refuse this boundary and the erasure of Karen people and sovereignty it enacts.

Foods from home made their way through the jungle on the Karen side of the border to students in the camp, carried by a network of people traversing this route or by students themselves when returning from visits home. These foods connected my young friends to the bodies of their parents and to their longed-for landscapes of home. Seremetakis (2017) theorizes commensality and memory in the face of austerity and late modernity via care packages of food sent to Greek university students from their homes in different parts of the country, with different regional food products and culinary special-ties. Like the students Seremetakis describes, my young friends in the camp enjoyed these familiar foods immensely when shared among themselves. Rather than being eaten alone in one's shoebox-like room, they were con-sumed in the company of others, sometimes including me.

In this context, foods from home prompted the remembering and re-telling of stories about farming, foraging, fishing, and caring for animals. These instances of sensory, embodied storying were similar to moments in a Georgian refugee camp described by Dunn (2017, 97–99). These stories highlighted the social relationships and feelings of joy and freedom imbued in performing such activities in the company of others. They revived in the minds of students forms of labor connected to these foods—labor that was not necessarily categorized as "work," as theorized by Tsing (2015, 77). These foods evoked more-than-human caretaking relations that persisted in stu-dents' home villages. This was even as students frequently remarked that they did not miss the hard, sweaty labor of farming.

A young friend of mine, Paw Wa, a woman in her midtwenties with frizzy curly hair and a soft but animated way of speaking, told me one day what it was like to forage and fish in her home village. She explained, "When we are foraging in the forest with our friends, or when we are fishing in the river, we get fresh air—and we feel so much joy!" In the realm of Paw Wa's vibrant storytelling, even the daily drudgery of walking the long distances to bring food to "mother's pigs" took on a humorous and nostalgic sheen. This was as opposed to feeding "father's pigs," located close to the house. The discrep-ancy in distance resulted in hotly contested debates about division of labor among her and her siblings, which she smiled and laughed telling me about.

Memory-laden foods from home that students received and shared in the camp prompted them to recount vivid stories of their villages and agricul-tural and foraging practices there. Such foods from home included laphet tho (pickled tea) and g'ney saw (honey). As the students emphasized, laphet tho was not just *any* pickled tea, but the pickled tea made by their family member, thus having a distinct recipe and taste. Likewise, the honey was not *just* honey, but "pure" wild honey captured from bees in the forest of Kawthoolei. They remarked on the wonderful distinct taste that g'ney saw from Kawthoolei has and told me that it could not be touched in comparison with any kind of store-bought honey. Not only were these foods, they were also medicines for the body and spirit (see Gordillo 2004). They brought strength to students who had lived for years far from home and without seeing their families. Often students were not even able to make a phone call to their parents because the places they came from still did not have cell signal. In the absence of other forms of connection, foods that moved across borders were carriers of "traveling memories" (Fiddian-Qasmiyeh 2013). They provided comfort to students who longed to go back to their "native places" and to the people they loved.

Naw Thi Klo Poe was a plucky student who often came to visit me and my friends Sema and Ruth in the female teachers' dormitory. One evening, Naw Thi Klo Poe came to our dormitory with a plastic jar in her hand. She said, "I brought something for you! Here, try this." She then made her way over to the kitchen counter, built from a piece of plywood, and began unscrewing the top of the container and mixing it with the leftover rice from our dinner. She brought a dish over to where Sema, Ruth, and I sat. It was a plate of pickled tea with rice. "This is from my home," she said. "My mother sent it to me. Here, have some."

Knowing how special this food must be to her, I protested, saying, "You can't just give it to us! I know how much it must mean to have food from home." "Yes," she said, smiling, "it tastes like home." Nevertheless she insisted, "I want you to try some. My mom makes it good."

Overwhelmed by her generosity, I took a bit in my spoon and tried it. It truly was wonderful. A rich pickled tea aroma filled my nose, and the pungent notes hit my tongue with their acidic punch of flavor. More significant than being an especially delicious version of a dish I had learned to love, however, was knowing that *this* pickled tea had been prepared by the hands of Naw Thi Klo Poe's mother, whom she sees only once a year over the school holiday and whom she was separated from for several years after coming to the camp.

Naw Thi Klo Poe's words about how comforting it was to taste the flavors of home resonate with Seremetakis's description of the importance of food parcels from home for Greek students studying in Athens. When students receive these parcels, they immediately tear them open to share with their friends. Seremetakis theorizes such organic moments as havens of commensality within the context of neoliberal austerity in Greece, including what has been described as "social cannibalism" (Arampatzi 2017). Similarly, for young people in Mae La, this space of communal enjoyment, drawn out by food from home, represents a sanctuary of memory (Nazarea and Gagnon 2021, introduction) within the increasing individuation of life brought about by the ethos of self-reliance imposed in the camp.

On a separate occasion, food from Naw Thi Klo Poe's home also emerged as significant. This time it brought new understandings for me about the place she came from and the values and affects tied to that place. It provided me with a window into the more-than-human relationships bound up with food, which in this case was also medicine: wild honey. In the early months of my time in the camp, I became sick and feverish and slept for the better part of three days.[8] During this time, Sema suggested I drink a remedy that she made with hot water, lime, and g'ney saw. The g'ney saw, she told me, was from Naw Thi Klo Poe's village.

The honey was in a slender plastic container with the logo of the NGO Save the Children on it. Ruth, who grew up in Yangon, asked about the veracity of the honey. She warned that "in Burma some people will try to pass off sugar water as honey. So, it's good to be sure." Sema reassured her that Naw Thi Klo Poe said it was "pure," that it had come from her village. Curious about this, I later asked Naw Thi Klo Poe about the honey and whether it had really come from her village. Her face lit up. "Yes!" she said, "My father caught the hive himself by climbing the tree and bringing it down in the middle of the night. It is wild honey from the forest. This is why it is so delicious. It is totally different from the honey you can buy at the shops in the camp, or even at Robinson's [the big, fancy supermarket in Mae Sot]." In that moment, honey from home surfaced as important not only for its prized flavor but also for its sensory recalling of the honey's procurement, which wove place and identity.

Strikingly, these are features shared with Elizabeth Dunn's account of the last spoonfuls of honey from home being shared and eaten, drunken and tearfully, by her interlocutor and friend Aleko. An ailing resident of a Georgian refugee camp, Aleko proclaimed with pride on bringing out the honey to share with her, "It's *mine*! It's from my bees!" (2017, 97). In both instances, honey—a portable and nonperishable food with flavors that are highly localized and seasonal because pollen comes from specific flowering trees and plants—crystalized sympoetic elements of placemaking that bound refugees, Naw Thi Klo Poe and Aleko, to the longed-for landscapes and social fabrics of their homes left behind. In this vein, Dunn recounts, "Eating the food that boomeranged back to the displaced—honey, jam, bottled fruit, and even homemade 'white lightning'—became a ritual in many cottages I visited. Almost always they were accompanied by beautiful, idealized and elegiac descriptions of the house and the land they had come from" (98).

Like Dunn, as well as Anastario, Seremetakis, Trapp, and others, I am interested in the ability of food to open up the political as it is tied to home and memory and as experienced through the senses. I am intrigued by the ways food allows critiques to be lodged that otherwise would be impossible to articulate because they hover at the realm of the affective—that is, as structural forces that are felt more than seen, acting on and between bodies (see Stewart 2007, introduction). Trapp (2016) discusses Rancière's notion of aesthetics as "the entire distribution of the sensible." Here "sensible" does not simply refer to reason or common sense but is a matter of perception: "what is perceived and perceivable" (433). Building on this, I read the movement of foods across borders, accompanied by their embedded sensory memories, as acts of refusal (Simpson 2014). These foods-transgressing-boundaries link fraught sanctuaries with edgy spaces of exile. In the process, foods from home and the students who receive them refuse the border between Thailand and Myanmar, which crosscuts Karen territory.

INTERLUDE: LATE-NIGHT PANCAKES AS METHOD

Naw Thi Klo Po, Paw Wa, and other students would come and visit me at my dormitory most evenings to drink tea and hang out. During these cherished visits with the students in the open-air kitchen, the pulsating sounds of cicadas would fill the velvety darkness, as massive stick-like insects circled the bare fluorescent bulb making thunking noises as they hit their exoskeletal bodies against it. We had to speak loudly to be heard over the din of the insects coupled with the deep, resonant croaks of the frogs coming from the creek just behind us, where men wearing LED headlamps waded with rolled-up pants and caught fish with their bare hands. These evenings, the students would sit around and chat with me, sometimes for hours until their eyelids began to droop, and they would finally say, "ni luh a gay, tharamu" (goodnight, teacher). As they departed around the corner, they would say a final "mi mu mu" (sweet dreams).

During these evenings I would sometimes cook pancakes for the students. Due to the central role of pancakes in my family rituals growing up—featuring in the tradition of Saturday "big breakfasts" in the sunlit kitchen—they hold special significance in my own sensory memories. They are, in many ways, one of my important "foods from home." Pancakes, it turned out, were also one of the few things I found I could make on the temperamental hot plate, with the one pan I had and the limited ingredients available to me. To my surprise, when I first made them for a group of students, they were a hit. They were a trendy food (pancakes and waffles featuring prominently in hip cafés around Chiang Mai) and warm and sweet enough to make a good late-night snack for the students at the end of their long days.

Taking a cue from Sema and Ruth, as well as the students, I had fallen into the habit of showering early in the evening, right after finishing my rice, to remove the stickiness of the day and feel fresh in the cooler evening air. This meant that I was usually in my floral print cotton pajamas, with an additional layer of floral print on top in the form of my favorite bathrobe. Thus, when they came to visit, students would find me in my florid camouflage, sipping hot tea and seated in front of my laptop catching up on fieldnotes or poring over student papers that needed grading, pen in hand. We would then playfully argue as I would try to get up to put water on to make tea for them, or to make pancakes. They would urge me to sit down and let them serve me instead. In this manner we went back and forth, expressing our mutual feelings of ana or being ana-day, a Burmese word that has to do with feeling ashamed by someone else's generosity, among other things, and which is notoriously difficult to translate.

On the occasions I won out, I found myself in the tower of the tharamu ga-lowa (white woman teacher's) dormitory, overlooking a darkened refugee camp filled with the sounds of night fishing, the croaking of frogs, and the whirring and thudding of insects. I found myself making pancakes in my pajamas to serve to Karen refugee college students who had become my friends. Their bright faces, laughter, and stories were among the things that saved me from total despair during those long months, in which I attempted to digest, and in the process partially internalize, the pain of incommensurable inequality and exclusion: the exclusion of people I had come to love from the realm of full humanity.

The Power of the In-Between

My emotional processing of the realities of life in the camp was distinct from how students understood themselves. Despite being seen by many people throughout the world as those most deserving of pity—as refugees living in a camp in the context of the world's longest civil war—my young friends did not see themselves in this light. Time and again the poetic justice of their words struck my ears when different students would exclaim, "Teacher, I feel *so* pity for you!" They would say this for various reasons, for example, because I was staying by myself, "only one," in an all-but-empty dormitory, or because I was far away from my home and my family. The students said this to me regularly, and with so much sincerity that I could not detect even a hint of irony. I noted the shock I felt at the reversal of our presumed emotional registers. (Shouldn't I be pitying them and not the other way around?) The shock I felt spoke more profoundly of my perception than theirs. The students were keenly aware of the injustices they faced in their lives. Yet, they did not see themselves as victims.

This was driven home for me one day while I was visiting one of my star students at her dormitory. Naw Do Ta Gay was so enthusiastic in class she would hardly let other students answer. As it turned out, she was equally earnest when telling me about her plans for the future. Chatting as she prepared dinner, and speaking so rapidly I was afraid she might run out of breath, she told me that her goal in life was to "become an international social worker." Her plan once becoming this kind of social worker, she said, was to go to Africa and "help the children in the refugee camps." After that, she said, she would continue her work in other parts of the world. She said this with all the starry-eyed self-assuredness of one of my freshman undergraduates at Syracuse University. The view in which she herself was that poor child in the refugee camp in need of help clearly did not occur to her. This is only one example of the countless times it became clear to me that my young friends in the camp saw themselves not as helpless victims, but rather as agential actors with definitive ideas about the kinds of change they wanted to make in the world.

In addition to viewing their oppression clearly while generally not being overcome by it, these young people also had potent insights into the movements of capital and lived experiences in different locations across the Karen diaspora. As a result of their position as border crossers between rural Karen villages and the camp, the students uniquely understood what was at stake as traditional subsistence livelihoods slipped into the background and as precarious forms of labor loomed large. Additionally, a window into the wider world of resettlement was available to them at the swipe of a screen. Decent 3G data was accessible in the camp, along with SIM cards sold in the camp market at rates inexplicably lower than at kiosks in Thailand. This provided them with all they needed to gain insight into resettled life in the various countries where Karen people now live.

The insightfulness of the students' global view from the camp became clear to me while observing the daily video chat habits of Naw Thi Klo Poe. She would regularly come to "visit" me in the afternoons. Living up to her membership in generation Z, this meant keeping me company with her physical presence while catching up with friends, relatives, and crushes scattered across the globe via Facebook video chats on her phone. This included talking with her aunts and cousins in Norway, as well as with friends now resettled in Australia, Canada, and the United States. After her chats she would talk with me, reflecting on what she had learned from her contacts about resettled life in these different countries.

One day, Naw Thi Klo Poe told me, "Some people really want to go to the U.S., but not me, *hu-uh*! I would NOT want to go there." This led to a long discussion in which she relayed to me much of what she had heard about how "people have to work so much in the U.S." She told me, "I heard that in the U.S., if you lose your job you will end up under the bridge." It took me a moment to understand what she meant by this. Then I realized that she was describing homelessness.

It struck me as profoundly paradoxical that this young woman, who had fled from her home and grown up stateless in a refugee camp, was feeling sorry for people who lived in one of the wealthiest, and supposedly most developed, countries in the world and proclaiming that she would not want to live there. "Maybe I would go to visit but *not* to live there," she continued. She showed me a staged YouTube video depicting Karen homeless people in the United States, something I learned had become a kind of mythology in the Karen diaspora. The phenomenon of homelessness seemed to shock Naw Thi Klo Poe, who was horrified at such a breakdown of social solidarity.

She then asked me whether it was true, as she had heard, that U.S. undergraduate students had to "pay so much money just to go to college," and whether it was true that they have a lot of debt. I told her that it was as bad as she had heard. She also showed me photos of gasoline receipts a Karen friend in the United States had photographed after filling up his pickup. It was a lot for a fill-up by U.S. standards, but it was a truly absurd amount of money by the standards of life in the camp. Naw Thi Klo Poe's friend had explained that he must pay nearly all this money just to go to work every day, and that he works six days a week in part to pay for it. This fact scandalized her.

Naw Thi Klo Poe's pointed analysis of resettlement in the United States, viewed from the camp, resonates with Retika Adhikari's findings that Nepali Bhutanese refugees awaiting resettlement anticipate and physically rest in preparation for resettled life in the United States. Her interlocutors observe that life in the United States will be marked by working "day and night," even before they have left the refugee camp. By doing so, she states, "refugees critically engaged the idea of American refuge . . . revealing that disillusionment with resettlement precedes their arrival in the U.S." (2021, 247).

Naw Thi Klo Po, informed by her experiences growing up and attending school on both sides of the border, had striking insights into the differences between life in Karen State and in the camp, as well as glimpses into resettlement. Her current position was a liminal one. She lived in Mae La for her studies but regularly contemplated aloud to me her yet-to-be-decided future, which might take her in one of several directions. From their in-between, and thus powerful, perspectives, Naw Thi Klo Poe and her peers delivered up pert judgments of social life in Burma, Thailand, and countries that make up the Karen diaspora. Examples included Naw Thi Klo Poe hilariously crossing her eyes and pretending to be a zombie when explaining to me what education under the Burmese government school system was like: notorious for rote learning and tamping down critical thinking.

Then there was the day in our sociology class when I taught a lesson on class and power. When I asked the students to apply the concepts to their lives, they exploded with recognition, providing critical insights and even making impromptu speeches. I was not prepared for the passion of their collective response. This response included students criticizing their confinement and unequal treatment in the camp as well as class difference in Myanmar. To illustrate this, they drew diagrams showing the power hierarchies of camp governance and social stratification in Myanmar as they understood them (figure 3.3).

At one point Saw Kay Thu stood up in the seat of his desk and gave an impassioned speech that reminded me of a scene from the nineties film

FIGURE 3.3 Students do class analysis of Myanmar society. Images courtesy of author's year three sociology students.

Dead Poets Society. He decried the injustice of their situation in the camp. He pointed out that Principal Jacob, the principal of the college, and camp section leaders could leave the camp anytime they wanted for trips to Mae Sot, the nearby town, but the students and other ordinary camp residents could not.[9] He even enlisted the other students in a call and response, asking them, "If Principal Jacob wants to go into Mae Sot right now, do you think he could?" "Yessss!" the class roared back. "Can we?" he asked, pointing at himself and the others. "Nooooo!" they responded in unison. That afternoon, in our packed, dimly lit classroom, their normally concealed rage at the total institution that kept them pinned rose quickly to the surface in emotive waves. Later, after the class, when I reflected on the lesson with Naw Thi Klo Poe, telling her how impressed I was with their response, she told me with a smile, "You asked us to talk about oppression. So, we thought to ourselves: we know about this. Now *this* is something that we can talk about."

Anthropologist Paul Stoller (2009) theorizes the liminal as a privileged space of insight and potentiality. He refers to this condition as "the power of the between." Just as it has been repeatedly invoked as the primary currency of anthropologists, the liminal perspective, or insider/outsider view, likewise affords refugee young people special understandings of transnational connections and possibilities. It also allows them privileged insight into the currents of global capital, with its concentric sites of intensity and abandonment (Dzenovska 2020). Here, I consider how such an edgy view, or "the point of view from disordered but productive edges—the seams of empire," as Tsing puts it (2012, 151), provides Karen young people in Mae La with a distilled awareness of the exploitation and oppression experienced by Karen people in multiple locations across home, exile, and diaspora.

They are educated about Karen history, language, and culture at schools in the camp. These are topics absent from the curriculum in Myanmar government schools (Yeo, Gagnon, and Thako 2020). I have heard a handful of claims from people working in development in Myanmar, most based in Yangon, that the so-called border Karen are misinformed or even "stuck in the past" when it comes to understanding the political reality of their country and their own people in Myanmar. On the contrary, I found that young people in the camp, particularly the mobile border-crossing youth whom I befriended, had keen insights into the broad array of possible futures facing them. They also voiced poignant reflections on life "back home." Their ability to reflect in such a way was perhaps aided by having some distance from those places and experiences. I suggest that such broad understandings of Karen peoples' situations on both sides of the Thai-Myanmar border and in diaspora are not readily available in the same ways to young people inside Myanmar or in diaspora. Neither are they readily accessible to older generations who have lived in the camp for decades.

Conclusion

Karen young people living in Mae La refugee camp refuse dispossession and alienation in creative ways through engagements with foods and plants tied to their homes. I argue that Karen young people living in the "between" (Stoller 2009) space of the camp preserved the sociality and biodiversity of their rural homes through the sensory-rich, embodied stories they told one another. These memory narratives were anchored in sensory engagements with food, plants, and landscapes of home. Through counter-acts and sensory memories, young people in the camp actively refused the erasure of their lifeworlds.

My young interlocutors' critiques of life in the camp were often articulated through the poetic language of sensory politics. Embedded in these critiques of food and other embodied aspects of life in the camp were macrolevel commentaries on the conditions of their lives. Such remarks laid blame at the feet of the wealthy and powerful: the UNHCR, the Myanmar government, the Thai government, and the leaders of wealthy countries of resettlement.

The flattening of more-than-human relationships playing out in the camp may seem to paint a discouraging picture of life for Karen refugees and their companion plant species (see Haraway 2008). Yet, at the same time, students in the camp remember and thus keep alive knowledge systems and love for biodiverse species from their homes. My young friends elaborately narrated memories of foraging, fishing, and farming in the company of friends and family. They dreamed aloud of the plants they would go and pick in the forest at that moment if they would not be punished by the Thai guards (the Palad) for doing so. (And, of course, some go anyway.) These same students also fought back against the biocultural simplification playing out in the camp by consuming foods from their home villages, including g'ney saw and laphet tho, while narrating the embodied delights connected to those foods.

In these ways, stories and food from home became a means of refusing the biocultural simplification taking place in the camp. Through such embodied connections with plants and foods, Karen young people reminded each other that a more biodiverse, convivial world still existed—and it was their home. In so doing, they challenged sweeping regional transformations that threatened to uproot them and alienate them from more-than-human caretaking relations that have historically made Karen Indigenous sovereignty possible.

Acknowledgments

Thank you to my friends and interlocutors in Mae La who made this research possible, and who took me in and provided me with hospitality. Thank you to my students, for coming to chat with me in the evenings to keep me from being lonely. I will never forget your kindness. Thank you to Winsome and Mhale, for making me laugh, and for all the meals spent together. Special thanks to Htee Lah and Tha Shee for your invaluable assistance with translation and making interview connections, as well as much conviviality, support, and good times throughout the research process. Thank you to the other teachers and administrators at the school in the camp, which must go unnamed here: you know who you are. I remain inspired by your dedication to sharing knowledge in difficult circumstances. Work on this chapter was supported by funding from the European Union's Horizon 2020 research and innovation program, under grant agreement No. 101079069 and by the Bringing Southeast Asia Home initiative at the University of North Carolina at Chapel Hill.

Notes

1. This is despite their also naming multiple benefits of life on the border compared to life in the village—the most important being the opportunity for education, which they did not have back home.

2. This has become even more the case after the Myanmar coup in 2021 and dramatically increased violence in Karen State, located in southeastern Myanmar.

3. The phrase "seeds of sovereignty" is inspired by and used in the spirit of the transformational work of the Seeding Sovereignty collective (seedingsovereignty.org).

4. The majority of these young people are not eligible for resettlement, having arrived in the camp after the cutoff date for resettlement in 2005, set by the UN High Commissioner for Refugees (UNHCR).

5. He is here referring to the principle of *nonrefoulement*, a precept of international law that forbids a country from returning asylum seekers to a country in which they would be in danger of persecution based on specific grounds. United Nations Human Rights Office of the High Commissioner, "The Principle of Non-refoulement Under International Human Rights Law," OHCHR, accessed March 9, 2024, https://www .ohchr.org/Documents/Issues/Migration/GlobalCompactMigration/ThePrinciple Non-RefoulementUnderInternationalHumanRightsLaw.pdf.

6. Many of the students I taught had come to the camp in the last five to ten years from homelands in Myanmar and Kawthoolei (Yeo, Gagnon, and Thako 2020).

7. For a similar context of nostalgic foods tied to left-behind home landscapes, see Diana Glazebrook's (2008) chapter on sago yearning among displaced West Papuans. As Glazebrook documents, longing for sago, and thus for home, can be a widely accepted cause of death.

8. In hindsight I understand that I was suffering largely from the psychosomatic effects of the death of my friend's baby and my visceral realization of the general hardships of life in the camp.

9. In fact, the students, who had very limited mobility in and out of camp, still had greater opportunities to leave the camp than those who were truly ordinary, impoverished camp residents. Such ordinary camp residents might not leave the small area of the camp for years on end.

References

Adhikari, Retika. 2021. "Temporalities of Resettlement: Date-Waiting for an American Future in a Bhutanese Refugee Camp in Nepal." *American Anthropologist* 123 (2): 237–49. https://doi.org/10.1111/aman.13537.

Anastario, Mike. 2019. *Parcels: Memories of Salvadoran Migration.* New Brunswick, N.J.: Rutgers University Press.

Arampatzi, Athina. 2017. "The Spatiality of Counter-Austerity Politics in Athens, Greece: Emergent 'Urban Solidarity Spaces.'" *Urban Studies* 54 (9): 2155–71. https://doi.org/10.1177/0042098016629311.

Dunn, Elizabeth C. 2017. *No Path Home: Humanitarian Camps and the Grief of Displacement.* Ithaca, N.Y.: Cornell University Press.

Dzenovska, Dace. 2020. "Emptiness." *American Ethnologist* 47 (1): 10–26. https://doi.org/10.1111/amet.12867.

Feuerverger, Grace. 2001. *Oasis of Dreams: Teaching and Learning Peace in a Jewish-Palestinian Village in Israel.* New York: Routledge.

Fiddian-Qasmiyeh, Elena. 2013. "The Inter-Generational Politics of 'Travelling Memories': Sahrawi Refugee Youth Remembering Home-Land and Home-Camp." *Journal of Intercultural Studies* 34 (6): 631–49. https://doi.org/10.1080/07256 868.2012.746170.

Glazebrook, Diana. 2008. "Unsated Sago Appetites." *Permissive Residents: West Papuan Refugees Living in Papua New Guinea,* 95–106. Canberra: ANU Press.

Gold, Ann Grodzins. 2009. "Tasteless Profits and Vexed Moralities: Assessments of the Present in Rural Rajasthan." *Journal of the Royal Anthropological Institute* 15 (2): 365–85. https://doi.org/10.1111/j.1467-9655.2009.01558.x.

Gordillo, Gastón. 2004. *Landscapes of Devils: Tensions of Place and Memory in the Argentinean Chaco.* Durham, N.C.: Duke University Press.

Haraway, Donna Jeanne. 2008. *When Species Meet.* Minneapolis: University of Minnesota Press.

Hromadzic, Azra. 2011. "Bathroom Mixing: Youth Negotiate Democratization in Postconflict Bosnia and Herzegovina." *PoLAR: Political and Legal Anthropology Review* 34 (2): 268–89. https://doi.org/10.1111/j.1555-2934.2011.01166.x.

Human Rights Watch. 2012. "Ad Hoc and Inadequate: Thailand's Treatment of Refugees and Asylum Seekers." Human Rights Watch, September. https://www.hrw.org/sites/default/files/reports/thailand0912.pdf.

Kantor, Hayden S. 2021. "People and Foods in Motion: Agricultural Dislocations and Culinary Remembrance in Bihar, India." In *Movable Gardens: Itineraries and Sanctuaries of Memory*, edited by Virginia Nazarea and Terese V. Gagnon, 59–84. Tucson: University of Arizona Press.

Nazarea, Virginia D. 2021. "Ontologies of Return: Terms of Endearment and Entanglements." In *Movable Gardens: Itineraries and Sanctuaries of Memory*, edited by Virginia Nazarea and Terese V. Gagnon, 253–72. Tucson: University of Arizona Press.

Nazarea, Virginia D., and Terese V. Gagnon, eds. 2021. *Movable Gardens: Itineraries and Sanctuaries of Memory*. Tucson: University of Arizona Press.

Norum, Roger, Mary Mostafanezhad, and Tani Sebro. 2016. "The Chronopolitics of Exile: Hope, Heterotemporality and NGO Economics Along the Thai–Burma Border." *Critique of Anthropology* 36 (1): 61–83. https://doi.org/10.1177/0308275X15617305.

Pongsawat, Pitch. 2007. "Border Partial Citizenship, Border Towns, and Thai-Myanmar Cross-Border Development: Case Studies at the Thai Border Towns." PhD diss., University of California, Berkeley.

Sassen, Saskia. 2010. "A Savage Sorting of Winners and Losers: Contemporary Versions of Primitive Accumulation." *Globalizations* 7 (1–2): 23–50. https://doi.org/10.1080/14747731003593091.

Sassen, Saskia. 2014. *Expulsions*. Cambridge, Mass.: Harvard University Press.

Seremetakis, C. Nadia (Constantina Nadia). 1994. *The Senses Still: Perception and Memory as Material Culture in Modernity*. Boulder, Colo.: Westview Press.

Seremetakis, C. Nadia (Constantina Nadia). 2017. "Sensory Antidotes in the Era of Crisis: The Symposium *Taste and Memory*." Lecture, American Anthropological Association Annual Meeting, Washington, D.C., November 2017.

Seremetakis, C. Nadia (Constantina Nadia). 2019. *Sensing the Everyday: Dialogues from Austerity Greece*. New York: Routledge.

Simpson, Audra. 2014. *Mohawk Interruptus: Political Life Across the Borders of Settler States*. Durham, N.C.: Duke University Press.

Simpson, Audra. 2017. "The Ruse of Consent and the Anatomy of 'Refusal': Cases from Indigenous North America and Australia." *Postcolonial Studies* 20 (1): 18–33. https://doi.org/10.1080/13688790.2017.1334283.

Standing, Guy. 2011. *The Precariat: The New Dangerous Class*. London: Bloomsbury Academic.

Stewart, Kathleen. 2007. *Ordinary Affects*. Durham, N.C.: Duke University Press.

Stoller, Paul. 2009. *The Power of the Between: An Anthropological Odyssey*. Chicago: University of Chicago Press.

Sutton, David. 2008. "A Tale of Easter Ovens: Food and Collective Memory." *Social Research* 75 (1): 157–80.

Sutton, David. 2021. "Revivifying Commensality: Eating, Politics and the Sensory Production of the Social." In *Movable Gardens: Itineraries and Sanctuaries of Memory*, edited by Virginia Nazarea and Terese V. Gagnon, 133–60. Tucson: University of Arizona Press.

Sutton, David, Nefissa Naguib, Leonidas Vournelis, and Maggie Dickinson. 2013. "Food and Contemporary Protest Movements." *Food, Culture, and Society* 16 (3): 345–66. https://doi.org/10.2752/175174413X13673466711642.

Thomas, Deborah A. 2019. *Political Life in the Wake of the Plantation: Sovereignty, Witnessing, Repair*. Durham, N.C.: Duke University Press.

Trapp, Micah M. 2016. "You-Will-Kill-Me-Beans: Taste and the Politics of Necessity in Humanitarian Aid." *Cultural Anthropology* 31 (3): 412–37. https://doi.org/10.14506/ca31.3.08.

Tsing, Anna. 2012. "Unruly Edges: Mushrooms as Companion Species: For Donna Haraway." *Environmental Humanities* 1 (1): 141–54. https://doi.org/10.1215/22011919-3610012.

Tsing, Anna Lowenhaupt. 2015. *The Mushroom at the End of the World: On the Possibility of Life in Capitalist Ruins*. Princeton, N.J.: Princeton University Press.

Yeo, Subin Sarah, Terese Gagnon, and Hayso Thako. 2020. "Schooling for a Stateless Nation: The Predicament of Education without Consensus for Karen Refugees on the Thailand-Myanmar Border." *Asian Journal of Peacebuilding* 8 (1): 29–55. https://doi.org/10.18588/202005.00a111.

CHAPTER FOUR

A Place with Diverse, Harmonious Knowledge

Banana Homegardens in Uganda

YASUAKI SATO

Bananas (*Musa* spp.) are one of the most popular fruits worldwide. They have a history of domestication and globalization, and different banana cultivation systems have been developed over the years (Kema and Drenth 2018; Komatsu et al. 2006, 77–119). These systems include two extremes: subsistence farming and plantation agriculture. The former is characterized by small scale, mixed cropping, genetic diversity, and Indigenous knowledge. In contrast, the latter is characterized by large-scale monoculture, few cultivars, and science and technology. These two contrasting systems reflect the contemporary world. In this chapter, I reconsider the significance of farmers' knowledge in the former system.

The Ganda people of central Uganda, East Africa, make banana gardens called *lusuku* (pl. *ensuku*) around their dwellings and manage them over generations. They consider banana the most important food crop and a crop with many uses. They spend much of their lifetime in their gardens and constantly watch over many kinds of bananas. By focusing on the people, their homegardens, and bananas, I aim to reveal the multifaceted nature of human-plant relationships (Galvin 2018, 233–49) and the virtue of farmers' expertise, which is different from scientific knowledge. In particular, I examine the folk taxonomy of bananas to understand the complexity of the farmers' relationships with bananas in their daily lives, a form of human-plant collaborative "refugia" in their homegardens.

The study was conducted intermittently from 2005 to 2018. Participant observation and interviews were used to obtain people's different perspectives on bananas. As a result of this study, I propose that farmers' homegardens are a place where different kinds of knowledge are harmoniously combined.

Banana and Its Cultivars: A Scientific Approach

Bananas have been diversified through complicated processes in the world. They are a vegetative and perennial herbaceous plant domesticated by selection from several wild species of the genus *Musa* of the Musaceae family (Stover and Simmonds 1987). Most banana plants grown as food crops are included in the section *Eumusa* and descend from either of the two wild species *Musa acuminata* and *Musa balbisiana*, or from both. An important characteristic of cultivated bananas is that they develop fruit by parthenocarpy, that is, without fertilization of ovules. They are thus propagated by vegetative means, not seeds. The genome types of cultivated bananas of the *Eumusa* section are denoted as AA, AB (diploids), AAA, AAB, or ABB (triploids), depending on the combination of genome types *Musa acuminata* (AA) and *Musa balbisiana* (BB). Cavendish is a well-known cultivar among those included in the subgroup triploid AAA. Companies involved in plantation agriculture export large amounts of these cultivars (JAICAF 2010).

In Uganda, so-called East African highland bananas (AAA-EA) are widely cultivated. They are classified as a subgroup of the autotriploid *acuminata* (AAA) and are distributed in humid areas in the East African highlands (the Great Lakes region of East Africa). This subgroup has traditionally been referred to as Lujugira-Mutika, because researcher Kenneth Shepherd (1957) placed these cultivars into two sets of botanical taxonomy: (1) Mutika, in which a bunch hangs down, the fingers are relatively large, and the apical part maintains a bottleneck form at maturity; and (2) Lujugira, in which the fingers are short, and the apex does not have a bottleneck form (JAICAF 2010; Karamura, Karamura, and Tinzaara 2012).

Evidently, after a few bananas (AAA-EA) were introduced into the region, they were separated into many cultivars by somatic mutation. Karamura (1999) and Karamura, Karamura, and Tinzaara (2012) identified 238 vernacular names of bananas (AAA-EA) growing in Uganda. After examining morphological variations and conducting a numerical analysis of these landraces, the researchers removed synonyms, leaving eighty-four distinct cultivars grouped into five clone sets: *nfuuka, musakala, nakabulu, nakitembe,* and *mbidde* (figure 4.1). The terms *cultivar* and *landrace* are distinguished

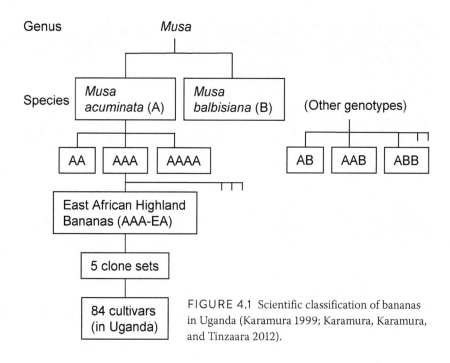

FIGURE 4.1 Scientific classification of bananas in Uganda (Karamura 1999; Karamura, Karamura, and Tinzaara 2012).

here. Cultivars are scientifically recognized and classified, while landraces are local varieties named and used by farmers, as I explore in this chapter.

Banana Cultivators and Their Homegardens in Uganda

In parts of the Great Lakes region in East Africa—including Uganda, Tanzania, Kenya, Rwanda, Burundi, and the eastern Democratic Republic of Congo—bananas are an indispensable staple food. People depend heavily on bananas for uses of agricultural landscape, subsistence economy, and local customs (Sato and Shigeta 2006, 405–11; Shigeta and Sato 2006, 413–20). In central Uganda in particular, the Baganda people maintain their identity as "banana eaters." Their primary staple food is the banana. They place a higher value on *matooke*, a favored banana dish (figure 4.2), than on other staple foods such as maize, cassava, sweet potato, and yam. In addition to being a staple food, bananas are also considered useful plants for various purposes, including in alcoholic beverages, sweets, and crafts.

The Baganda people have developed a unique livelihood system based on banana cultivation for hundreds of years (Sato 2011). Schoenbrun (1998) es-

FIGURE 4.2 *Matooke omunyige*, a dish of steamed and mashed bananas. Photo by author.

timated that the intensive cultivation of bananas expanded along the northern and western shores of Lake Victoria between the twelfth and fifteenth centuries. The Baganda people are a majority group in Uganda, with a population of about four million in 2002 (UBOS 2005). They belong to more than fifty patrilineal clans. The Buganda Kingdom ceased to be a modern nation in 1966. Today, Buganda means central Uganda, the territory of the Baganda people. The present study was conducted in a village of Kirumba Subcounty, Kyotera County, Rakai District in Buganda, at the time of the survey.

The Baganda people's interdependence with bananas has created a unique landscape of a forest of bananas. One small forest is set with one dwelling, and a series of countless small forests make up a large forest. Although at first glance the bananas seem to have been randomly planted and left to grow on their own, the area is in fact well organized and taken care of by the people (figure 4.3). In their forest, or in other words, their homegardens, bananas and other staples, leafy greens, and fruits are planted, allowing people to live autonomously as if they were in a safety net. One elderly man told me the following story: During the civil war in the 1980s, there was a

FIGURE 4.3 Management of *lusuku*, a homegarden of bananas. Photo by author.

period when he could not go out because of the lack of security. There was a time when he survived for months on only the food harvested from his homegarden.

In Luganda, a Bantu language spoken by the Baganda people, no word directly corresponds to "homegarden." A field or garden in general is called *enimiro*, but only a banana garden is especially called *lusuku*, not *enimiro*. Since most banana gardens surround dwellings, *lusuku* has the equivalent meaning of a homegarden. Figure 4.4 shows a map of an informant's premises, including his homegarden. Each dwelling is surrounded by a homegarden of bananas, which is carefully managed all year round. Different landraces are planted according to the site because soil fertility and accessibility vary from place to place within a garden. Furthermore, homegardens are used not only for food production but also for housework and rest. It is also common to find the graves of the residents' ancestors in these homegardens. People in Uganda maintain many landraces of bananas on a small scale (Gold et al. 2002, 39–50). According to the sample households in the research site, each household has an average of approximately twenty landraces. Of these, the

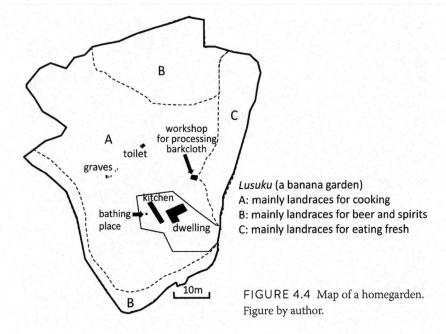

FIGURE 4.4 Map of a homegarden.
Figure by author.

average number of staple food landraces is about ten. In other words, each homegarden is a living storehouse of banana genes.

A homegarden of bananas is not only a bountiful blessing from nature for them, but also a place that contains various feelings of desire and envy. When I visited one family's homegarden, which is well managed, the owner told me that the condition of a banana garden indicates the condition of the family. He had the following to say: "This banana garden is special for me. I don't allow children to play there, and I never allow strangers to enter without permission. I have to be careful because there are banana thieves these days."

His words convey his pride in his garden and the seriousness with which he manages it. In this area, there is a custom for a man to open a banana garden when he becomes independent from his parents, and to take in a wife. When choosing a plot of land, he considered whether the soil would grow bananas well. After opening the garden, he managed it to provide food for his family for decades. In other words, he has lived his life with his homegarden.

His wife is responsible for harvesting the fruit when it is ready. I observed curious behaviors of hers. One was to leave one of the harvested bananas in its place. She said that this means keeping the banana garden from being completely harvested. It is thought to be a kind of offering to the gods. The second is to pick a leaf from a certain weed (*Phyllanthus amarus*) and hook it onto a nearby banana plant. She said, "This grass makes the bananas grow better. I hook the bananas after I harvest them so that we can get more of them." Third, when cutting the remaining part of the plant with a knife after harvesting, she removed the flower and hid it under the leaves laid on the ground. If a witch doctor could get it, he could cast a spell on this garden so that it would not produce bananas at all and instead fill another garden with bananas that he had spotted, she said. Thus, their homegarden is a place that should be kept clean, treated humbly, and protected.

Folk Taxonomy and Cultural Perception

The Baganda people can perceive the multifaceted nature of bananas, almost as if they can talk to bananas. Luganda has a rich vocabulary of terms for bananas. Various nouns are used to refer to the different parts, forms, and applications. For instance, a leaf sheath, or a pseudostem, is called *olugogo*, a leaf is called *olulagala*, and the last leaf is called *engozi*, which also denotes the cloth used for carrying a baby on one's back. The leaves used for wrapping are called *oluwumbo*, and the covering leaf is called *essaaniiko*. These last two are used for cooking the steamed dish *matooke*.

When considering the folk taxonomy of bananas, it is important to regard the existence of words that describe all kinds of bananas. The word *banana* in English is used in all contexts and to refer to every part of the plant. In Luganda, it is difficult to find a word that perfectly corresponds to it. The noun stem -*tooke*, however, is commonly used to refer to the plant, the finger, and a processed banana dish. The banana plant is called *kitooke*, which includes all banana landraces. For instance, an attractive banana plant is called *kitooke kirungi*, while an unattractive one is called *kitooke kibi*. People distinguish between different banana plants based on their characteristics. In addition, sociocultural contexts influence people's perspectives. Through interviews, tests of knowledge, and daily observations, four kinds of taxonomies or perceptions were identified: (1) large categories (composed of landraces), (2) landraces, (3) genealogies, and (4) individual plants.

Large Categories
Landraces can be grouped according to the pseudostem color. The Baganda people divide all banana landraces into two categories based on the pseu-

dostem. The first is called *kitooke kiganda* (our original banana). The land-races under this category are believed to have been brought to Buganda in ancient times. They have many black dots on the pseudostem. The second category includes landraces that have only a few black dots on the pseu-dostem. The landraces under this category are believed to have arrived in Buganda relatively recently.

Landraces are also categorized according to their use. Based on use, *kito-oke* are divided into two groups. The first is called *kitooke kyemmere* (staple food bananas) or *matooke*. Several plants in this category are called *enkota yetooke*. The second category is *mbidde*. These landraces are used for brew-ing beer and spirits or for making light meals. The sap from the pulp of some of these landraces is too sticky for cooking. *Matooke* is traditionally in the female domain, while *mbidde* is in the male domain. Because all bananas in this area can be used as food, every banana plant is grouped into one of the two categories. The above classifications based on pseudostem color and uses are utilized all over Buganda and are in accord with botanical clas-sification. A blackish pseudostem is also a key characteristic of AAA-EA. Landraces suitable for cooking correspond to all clone sets, except for the *mbidde* clone set.

Landraces

Bananas are also grouped into dozens of landraces called *ekika* (a type or variety) or *ekika kya kitooke* (a landrace of bananas) by shape or growth characteristics. Each landrace is thought to have distinct characteristics of appearance, growth rate, and disease tolerance. Both large categories and landraces are considered to have similar characteristics, even after updates in generations of plants or transplanting. The pseudostem color is affected by soil conditions, and some recently introduced cultivars are used for both cooked food and beer. Nonetheless, the taxonomy seems quite stable.

To understand the Baganda people's perception of landraces, I collected data on the etymology of landraces. Table 4.1 shows the information I ob-tained from some elderly people in the village. The table contains a much smaller number than the actual number of cultivated landraces, because it has been many years since a landrace was named. Some informants associ-ated similar words with a named landrace on the spot. Many of the names listed in table 4.1 refer to the appearance of plants, especially their bunches and fingers. The female prefix *na-* is used before many of the named landraces. Matovu and Byangwa (1995) indicated that the structure of a typical banana landrace name is as follows: "female prefix + class prefix + root." This is similar to the structure of a female person's name, suggesting that bananas have a feminine image.

Banana plants are grouped into landraces according to different combinations of characteristics. Their identification is complicated, however, for both farmers and natural scientists. The informants use the following characteristics of a banana plant to distinguish landraces:

1. Whole plant: length and figure of the plant
2. Leaves: (a) length, size, width, numbers of leaves, and stickiness of sap from leaves; (b) color of petiole
3. Fruit: (a) length and size of bunch; (b) number, size, and compactness of hand; (c) length, size, compactness of fingers, and stickiness of sap from fingers
4. Pseudostem: color of pseudostem
5. Suckers: number and figure of suckers

Of the twenty characteristics, eighteen are related to the plant's appearance. For grouping landraces according to leaves, the Baganda people use four pairs of antonyms: *-nene* (big) and *-tono* (small), *-wanvu* (long) and *-inpi* (short), *-gazi* (wide) and *-fumba* (slender), and *-ngi* (many) and *-tono* (few). Regarding bunches, they use the terms *-gasse* (compact) and *-banga* (scarce) to categorize landraces. In addition to appearance, they also look at the growth speed and taste of the banana.

The five groups of characteristics also suggest that many are contiguous. Each landrace is distinguished not only based on the dichotomy of yes/no or included/excluded but also on several other dimensions. This complicates their identification. Informants were asked to explain the relative order of landraces. As a trial, I gave two informants thirty-five cards with one landrace name written on each of them. I then asked the informants to classify them according to the length of leaves and fingers. One informant classified them into thirteen levels of leaves and twenty-one levels of fingers. The other informant classified them into eight levels of leaves. There are thus significant differences in such identification skills among people.

Genetically, these characteristics may reflect quantitative traits. Natural scientists have not completely clarified the synonyms of a large number of

TABLE 4.1 Etymology of banana landraces

Landrace	Etymology	Notes on characteristics
Nnandigobe	Before people started living there, the king of Buganda was apparently a snake. Kintu, the first king of human beings, got rid of it. *Naligobwe* = run after	The bunch bends in one direction.
Nakabululu	*Kabululu* = low -*webulala* = to be short and fat	The bunch is compact. It is difficult to separate individual bananas from a hand of bananas. The plant is the shortest among all landraces.
Nakyetengu	*Kyetegulakyokka* = a dwarf	The plant is the shortest among all banana landraces.
Kiriga	*Endiga* = sheep	
Nalukira	*Lukiro* = a tail	
Naluwezinga	-*ezinga* = to spiral	
Mbwazirume	A dog takes a bite of something. *Mbwa* = a dog	The petioles of leaves are red.
Nakinyika	Hard fruits that are soaked in water before cooking	
Nsakala	*Kisagala* = big	There are gaps between the hands of bananas.
Nakitembe	Kintu, the first king of human beings in Buganda	As in *nsakala*, there are gaps between the hands. The hands are shorter. It is used in ceremonies for the birth of twins and for bereaved families in mourning.
Bogoya	Unknown	The pseudostem is green. The fingers are long.

Name	Meaning	Description
Muvubo	*Muvubo* = a pipe -*vuubiika* = to take something in one's mouth	It is similar to *nsakala*. The bunch is shorter, and the fingers are long and large.
Musa	Unknown	The pseudostem is green. It is used for making beer. During a drought, it is consumed as food.
Nsowe	Unknown	It brings many suckers.
Salalugazi	Unknown	The leaves are long.
Kabula	Unknown	It is used to make beer. It is easily confused with *kibuzi* (a staple landrace).
Nngomba	*Mugumba* = a woman who cannot bear children	The pseudostem is extremely black. It bears a few suckers. It is used to make beer.
Kayinja	It makes alcohol as strong as stone. *Jinja* = a stone	
Kibuzi	*Enbuzi* = a goat -*kibura* = to be too small to be visible	
Mukubakonde	*Ekikoola* = a fist	The bunch looks like a fist.
Njoya	*Njoya* = I need to eat	
Butobe	-*tobeka* = to be compact	The fingers are compact.
Nabussa	There is no food because one does not work. *Trilirilebusa* = no business *Busa* = nothing to exchange	

landraces in the East African highlands. Karamura (1999) asserted that local knowledge can be used as a starting point for further clarification. The Baganda people have different knowledges of banana landraces. This may be because of differences in their experience and environment. To determine the variation in the community's knowledge, farmers' banana landrace identification skills were tested in a household's homegarden.

Participants were thirty villagers (sixteen males, fourteen females) with a wide age range from adolescents to octogenarians. Each participant was asked the name of a landrace while in front of that plant. There were twenty-five plants in total (eight with bunches, and seventeen without bunches), and they belonged to the category *kitooke kiganda*. Each was a different landrace. Scoring was based on the number of answers that matched those of the homegarden owner, the head of the household. The results of the test with all twenty-five plants are shown in figure 4.5a, and the results of the test with only the eight plants with bunches are shown in figure 4.5b. A score of twenty-five on the former and eight on the latter meant that all answers matched those of the garden owner.

It is noteworthy that all participants found the test difficult. Figure 4.5a shows that the highest score was sixteen out of a possible twenty-five. No participant gave the correct answer for six of the plants. Even the garden owner's wife, who manages the garden daily, gave some answers that differed from those of her husband. The wife explained that her husband was the one who transplanted most of the banana plants in their homegarden, which is why he knew more. Furthermore, identifying landraces of banana plants without bunches is significantly more difficult than those with fruit because the flowering part is vital for identification.

Most of the villagers believe that women can distinguish the plants better than men. Figures 4.5a and 4.5b show that women tend to distinguish landraces better than men. Women took less time to answer than men. Moreover, there was no clear correlation between participants' age and scores. Such knowledge does not seem to directly correspond to the length of experience. There were clear differences, however, in landrace identification skills between the under-ten and over-ten age groups. Many children refused to participate in the test because of their inadequate knowledge. Several respondents explained that children in the region start helping their parents with tasks such as harvesting the bunches and cutting banana leaves for cooking from about ten years of age, which is when they start to learn how to

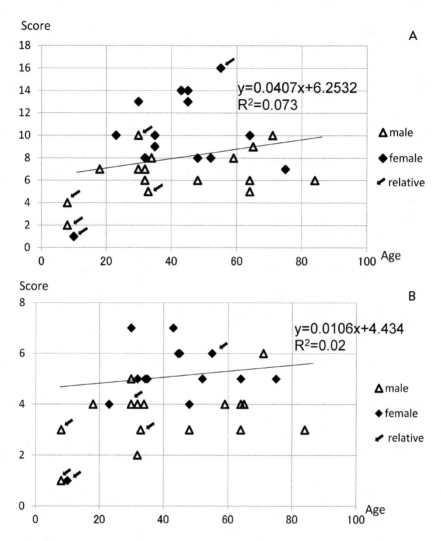

FIGURE 4.5 (a) Test scores for landrace identification of twenty-five banana plants; and (b) test scores for landrace identification of eight banana plants with bunches. Figures by author.

identify different landraces. The small deviation in the scores indicates that some knowledge is widely shared in the community. For example, fifteen of the sixteen men and thirteen of the fourteen women identified *kibuzi*. Furthermore, fourteen of the men and all the women identified *nsakala*. Both *kibuzi* and *nsakala* are popular varieties.

Genealogy

The Baganda farmers consistently look after the banana plants in their homegardens for long periods. Even after updates in generations of banana plants, they are treated continuously. This reflects the way people recognize banana plants in their homegardens. Luganda vocabulary includes the term *ekikolo* (pl. *ebikolo*), which is a unit for counting banana plants. It has a broader meaning than "a shoot" and is similar to "a mat with connected shoots and suckers." *Ekikolo* also includes banana shoots that are divided naturally and change their positions in the process of successive generations. Before they are transplanted, the shoots are called *ekikolo* (one group). In other words, the Baganda people are aware of the genealogy of banana plants.

Part of a map of a garden, drawn with the help of an informant's recollections, is shown in figure 4.6. The informant bought this land in 1970 and managed it for about thirty-five years. When he was interviewed, he recalled the positions of almost all the banana plants he had planted. His

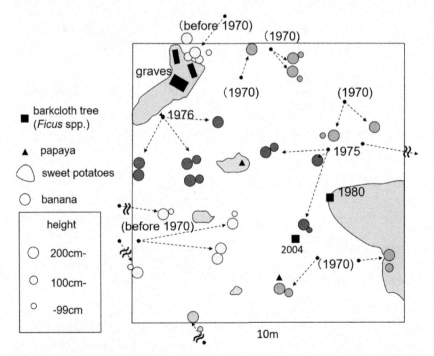

FIGURE 4.6 Positions of banana plants planted and moved over generations of plants based on an informant's recollections (years indicate the year of planting, and arrows indicate movement through successive generations). Figure by author.

related responses during the first and second interviews, separated by a few months, were almost the same. In the figure, the banana plants linked by arrows are called *ekikolo*. These plants include those planted before 1970, in 1970 when he bought the land, and after 1970. Although he could recall the links between the initial and current positions of the plants, he had only vague memories of how many generations were involved in the process. Such knowledge, which depends on people's individual memories, differs from the widely shared knowledge of landraces. For example, his wife remembered the positions of *ekikolo* quite differently from her husband. This is because, according to her, she did not have much planting experience.

Individual Plants

Regarding genealogical knowledge, many of the Baganda people can recall the landrace and positions of individual plants in their gardens. When indicating an individual plant, they say, for example, *litudde ebugwanjiba* (*malaalo, ekubo*), "the banana plant whose flower droops toward the west (the grave, the road)," or *matooke abiri ga kibuzi temako eryamanga* (two bunches of *kibuzi*, a landrace grown at the bottom of the slope). They do not assign individual names to their banana plants as people name their pets; rather, they remember individual plants by relating them to specific positions. If a landrace is rare, then its name is used as a proper name.

In addition to landrace plus position, the Baganda people also recognize individual banana plants in connection with important past events or through their appearance in someone's dreams. First, it is customary in Buganda to plant trees or crops in homegardens as a remembrance of events. People plant bananas, coffee, mango, bark cloth trees, and other crops to observe events such as the new year, birth of children or grandchildren, death of relatives, construction of new dwellings, famine, and other occurrences. This custom is called *ekijukizo*. Looking at these plants helps people recall past events. Interestingly, no strict rules or norms govern the commemoration of these events and the placement of plants. This custom is followed for both positive and negative experiences, and people freely choose where they want to place the plant. Generally, they plant beside or near their houses. They are also free to select any kind of plants, although plants that grow for years are considered better. People often plant bananas because they think that banana plants are continuously updated. After planting, these plants for remembrance are taken care of in the same way as other plants. An older person I interviewed said, "In the past, we did not have anything to write on, so we used to plant trees to remember our events." Even today, this remains a popular custom, and people associate individual plants in their homegardens with their experiences.

People also associate individual banana plants with their dreams (*eti-rooto*). In the interviews, informants explained that the landrace and position of the plants in their dreams matched their actual landrace and position. This suggests that their experiences with each plant in their homegardens are reflected in their dreams. For example, an informant narrated a dream he had had one day: "In my dream, the leaves of *nakyinika* [landrace] in my garden withered because of *kayovu* [a kind of pest]. These banana plants produced only small bunches. I peeled the outer part of the pseudostems to prevent further damage, but the plants continued to wither. After waking from the dream, I went to the same place in my garden as in the dream. There were some *nakinyika* plants. I removed them and planted different landraces such as *nsakala* and *nakitembe*. I might have had this dream because I used to always keep my eye on *nakinyika*, but it still grew poorly." He also talked about another dream regarding individual trees in his garden for making bark cloth. These examples suggest that people associate banana plants as well as other trees in their gardens with their dreams.

Plasticity and Variability of Knowledge on Landraces

In the previous section, I explained the folk taxonomy of bananas and people's recognition of them in terms of large categories, landraces, genealogy, and individual plants, each accompanied by specific context and examples. In this section, I focus on the dynamic aspect of people's knowledge concerning (1) the subdivision of landraces, (2) the transformation of landraces, and (3) unknown plants. I also consider the plasticity and variability of associated knowledge of landraces.

Subdivision of Landraces

Among the four levels discussed in the previous section, plants under large categories and landraces are named. There is no naming according to genealogy, however, and individual plants are also not named. In addition, there is further classification and occasional naming of landraces. In one interview, a sixty-five-year-old woman shared her knowledge of the subdivision of the landrace *nakabululu*. After her divorce, she returned to her parents' home in 1972 and took over their homegarden, continuing to grow the same land-races as they had. She informed me that there are two types of *nakabululu*, a major landrace. One type has many small fingers, which is what most people know about, while the other type, which has larger but fewer fingers, exists only in her garden. This is what her parents had told her. She continues to grow this type without naming it.

In another interview, a fifty-two-year-old woman mentioned that there are two types of *nakitembe*, another major landrace. One type is called *na-kitembe omuganda* (our original *nakitembe*), which has a whitish midrib and large fingers. The other type is called *nakitembe omusese* (*nakitembe* of the Sese Islands in Lake Victoria), which has a blackish midrib and smaller fingers. When she married and moved to her husband's house, she found these plants in his garden. She claimed that only the household members knew about it.

Furthermore, *gonja* is a landrace classified as *mbidde* (beer), which is peeled and roasted before eating. Botanically, it is classified as AAB plantain, and its shape is quite different from that of other bananas. Most villagers know that *gonja* has different types, and some people grow several of these types. Some are named using the same name followed by another word: *gonja muganda, gonja manjaaya, gonja owamasansabari, gonja nakakongo*, and *gonja mamba* (as recalled by informants). Yet, I often heard informants say, for example, "I do not know the names of *gonja* types, but I grow two types of *gonja*" or "*gonja* with few hands and large fingers." These examples refute the argument that a landrace name must correspond to a seedling, and that they must be introduced as a set. The Baganda people continue to grow even unnamed plants.

Transformation of Landraces

People in the study village know that landrace names continue to be used even after generations of plants are updated. At the same time, they also know that certain landraces change to other landraces. *Mbwazirume* and *nakitembe* are popular landraces. Most villagers know that *mbwazirume* changes to *nakitembe* irreversibly, but each one has different knowledge about it. Mr. A said that there is only a change in the pseudostem color, while Mr. B asserted that the bunch and leaves become smaller in addition to changes in the pseudostem color. Ms. C mentioned that the bunch maintains the same form and size, but the fingers become smaller. Further, Mr. D stated that the *nakitembe* that is transformed from *mbwazirume* is in fact different from the original *nakitembe*, although both are called *nakitembe*. People thus carefully compare such exceptions with their existing knowledge. They generally do not share such information with each other. Instead, they acquire their own experience and knowledge. Their homegardens are places where such acquisition takes place.

Unknown Plants

Some plants do not fit any type of landrace. Here, I discuss how the Baganda people react to such cases. An informant shared a story from 1984: A strange-shaped banana plant was found beside a road at the site of a household in a neighboring village. Four bunches and six flowers grew out of one shoot of this plant. Four sets of bunches and a male bud were connected to a peduncle. Its height was small—about 2–2.5 meters. Two male buds were attached to the bottom of the peduncle. Many people visited the site to observe the plant. Some came from distant places, as far as forty kilometers. Because many people started visiting the site, the owner of the plant put up a fence to prevent trespassers. Opinions were divided about whether the strange plant was a sign of fortune or misfortune. People who considered it fortunate paid attention to its characteristic of bearing many fruits. To use it as magical medicine, many people climbed over the fence, cut a part of the plant, and took it away with them. Perhaps, therefore, the plant eventually died. It was not given a new name.

Another villager shared a story from 2005 about the chairperson of a neighboring village who received a seedling of a new variety of banana. The chairperson's friend who gave him the seedling said that it had been originally introduced by the Ugandan government, had a large shape, and produced large bunches. When the chairperson planted and grew the seedling, to his surprise, he found that it did not bear any fruit. When I asked whether he would continue to grow it, he answered that he would observe it. In 2006, when I visited again, he mentioned that the next generation of the plant also did not bear any fruit. Some informants said that a banana plant that does not produce fruit is called *mukunga* or *kisulaani*. *Mukunga* is used to refer to a woman who cannot produce milk, while *kisulaani* means "misfortune." Although the chairperson kept growing this variety, he never named it. These examples indicate that the Baganda people rarely name strange or new plants or share them with others. Instead, they tend to keep such plants to themselves and simply continue to closely observe them.

What Factors Help Harmonize People's Diverse Knowledges?

In this chapter, I have examined the Baganda people's perspectives on the diversity of bananas present in the banana-based livelihood system in Uganda. From an ethnobotanical approach, I found four levels at which people categorize or distinguish banana plants: (1) large categories (composed of landraces), (2) landraces, (3) genealogy, and (4) individual plants. A detailed examination revealed that this knowledge is both widely shared by farmers and based on personal elements of individual farmers. In other words, such knowledge includes both formal and static knowledge, as well as personal experience. The former is common all over Buganda and is related to the functions of the landraces, such as their use as food. This kind of knowledge is also in accord with scientific knowledge. By contrast, the latter kind of knowledge depends on individual farmers' experience in cultivation. It is deepened separately and shared only with close relatives and friends. Such knowledge also enables people to view bananas from multifaceted and diverse perspectives.

Regarding the plasticity and variability of such knowledges, people are tolerant of separate knowledge, ambiguity in their knowledge, and the coexistence of different kinds of knowledge. When new experiences contradict existing knowledge systems, they are reluctant to revise their existing knowledge and instead accept the overlapping of new information with existing knowledge. The examples presented here indicate the Baganda people's attitude of continuous observation of reality, or what Anna Tsing describes as "arts of noticing" (Tsing 2015, 25, 37).

Such knowledge and tendency of the Baganda people can be considered from the perspective of the materiality of bananas as companion species (Galvin 2018, 233–49; Haraway 2007). Botanically, the phenotypic diversity of bananas, especially AAA-EA, is so intricate that it is difficult to morphologically identify cultivars. Both scientists and farmers face some difficulty in identifying them. On the other hand, banana plants have large and distinctive visual appearances, which can promote detailed consideration. This feature of bananas may also encourage people to be tolerant of knowledge diversity and to continue to seek knowledge to understand the nature of banana plants, as if they can engage in a dialogue with the plants.

The homegarden system in Buganda contributes to the people's attitude of embracing ambiguity and contradiction. As they maintain soil fertility in their homegardens, they use the land for decades without shifting. They can associate each banana plant with a particular position in their homegardens. Moreover, they can observe, experiment with, and compare the plants there. For them, homegardens are places where various customs and feelings are embedded. Thus, the garden is a unique place where diverse practices and memories are embraced. Regarding the dichotomy between clarity and ambiguity, Nazarea (1998) highlighted it in the conservation of sweet potato varieties in the Philippines. She insists that "the problem is when fuzzy, adaptive local perceptions that foster diversity collide with the ordering, reductionistic principles of formal science, be it agronomic or economic in nature" (1998, 72). In the case of bananas in Buganda, multiple perspectives of farmers, characteristics of bananas, and homegarden systems ensure the mutuality of human-banana relationships as refugia in a genetic and cultural sense, helping resolve the conflict between clarity and ambiguity.

References

Galvin, Shaila S. 2018. "Interspecies Relations and Agrarian Worlds." *Annual Review of Anthropology* 47:233–49. https://doi.org/10.1146/annurev-anthro-102317-050232.

Gold, Clifford S., Andrew Kiggundu, A. M. K. Abera, and Deborah A. Karamura. 2002. "Diversity, Distribution and Farmer Preference of *Musa* Cultivars in Uganda." *Experimental Agriculture* 38 (1): 39–50. https://doi.org/10.1017/S0014479702000145.

Haraway, Donna J. 2007. *When Species Meet*. Minneapolis: University of Minnesota Press.

JAICAF (Japan Association for International Collaboration of Agriculture and Forestry), ed. 2010. *Cooking Banana in Africa*. Tokyo: JAICAF.

Karamura, Deborah A. 1999. "Numerical Taxonomic Studies of the East African Highland Bananas (*Musa* AAA-East Africa) in Uganda." PhD diss., University of Reading and INIBAP.

Karamura, Deborah A., Eldad B. Karamura, and William Tinzaara, eds. 2012. *Banana Cultivar Names, Synonyms and Their Usage in East Africa*. Kampala: Biodiversity International.

Kema, Gert H. J., and André Drenth, eds. 2018. *Achieving Sustainable Cultivation of Bananas*. Vol. 1, *Cultivation Techniques*. Cambridge: Burleigh Dodds Science.

Komatsu, Kaori, Koichi Kitanishi, Satoshi Maruo, and Rosei Hanawa. 2006. "Comparative Study of Banana Farming Culture in Asia and Africa: The Diversity of the Cultivars." *Asian and African Area Studies* 6 (1): 77–119 (in Japanese).

Matovu, Kasalina, and N. Byangwa. 1995. "Environmental Conservation Through Cultural Practices and Language Use." In *Uganda: A Century of Existence*, edited by P. G. Okoth, M. Muranga, and E. Okello-Ogwang, 129–57. Kampala, Uganda: Fountain.

Nazarea, Virginia D. 1998. *Cultural Memory and Biodiversity*. Tucson: University of Arizona Press.

Sato, Yasuaki. 2011. *Life-World of Banana Cultivators in Uganda: An Ethnoscience Approach*. Kyoto, Japan: Shokado (in Japanese).

Sato, Yasuaki, and Masayoshi Shigeta. 2006. "Ethnobotanical Comparison of Banana and Enset Use in Africa." In *Proceedings of Kyoto Symposium: Crossing Disciplinary Boundaries and Re-visioning Area Studies, ASAFAS and CSEAS, Kyoto University*, edited by J. Maruyama, L. Wang, T. Fujikuru, and M. Ito, 405–11. Kyoto: Nakanishi.

Schoenbrun, David L. 1998. *A Green Place, a Good Place: Agrarian Change, Gender, and Social Identity in the Great Lakes Region to the 15th Century*. Oxford: James Currey.

Shepherd, Kenneth. 1957. "Banana Cultivars in East Africa." *Tropical Agriculture* 34 (4): 277–86.

Shigeta, Masayoshi, and Yasuaki Sato. 2006. "Ethnobotanical Comparison of Banana and Enset Agricultural Practices in Africa." In *Proceedings of Kyoto Symposium: Crossing Disciplinary Boundaries and Re-visioning Area Studies, ASAFAS and CSEAS, Kyoto University*, edited by J. Maruyama, L. Wang, T. Fujikuru, and M. Ito, 413–20. Kyoto, Japan: Nakanishi.

Stover, Robert H., and N. W. Simmonds. 1987. *Bananas*. Tropical Agriculture Series, 3rd ed. New York: Longman.

Tsing, Anna Lowenhaupt. 2015. *The Mushroom at the End of the World: On the Possibility of Life in Capitalist Ruins*. Princeton, N.J.: Princeton University Press.

UBOS (Uganda Bureau of Statistics). 2005. *2002 Uganda Population and Housing Census Main Report*. Kampala: UBOS.

PART II

Sovereignty

SIKASSO SNOW

never thought I
would see it here
in this Otherwhere,
no plantation in sight

no patterned pods
for the picking,
nor calloused hands
to plow & gin

an untouched December
bluff surrounded by scrubs of green
blowing along
a dust-whipped road heading south
toward no one's harvest.

—Fred L. Joiner

CHAPTER FIVE

Homegardens in the Southern Andes

Cultivating Agrobiodiversity, Learning, and
Sovereignty from Interculturality

JOSÉ TOMÁS IBARRA, JULIÁN CAVIEDES,
AND ANTONIA BARREAU

Next to each house, you will find the huerta. From a distance, this vibrant area, enclosed by a protective fence, resembles a diverse, multilayered forest more than a typical garden. As you approach, you can discern a distinct structure—rows etched into the soil and plants reaching for the sky, revealing careful cultivation and thoughtful management. Unfasten the modest gate, securely closed to prevent the entry of uninvited guests, and you will be greeted by a wide array of neatly organized plants: tall and short, displaying various shades of green and unexpected hues, adorned with flowers and budding fruits—some even bearing hanging pods. Birds gracefully dart among the scarecrows, while insects buzz between blossoms—it is a place teeming with life! We find ourselves in a homegarden, a lush and diverse social-ecological system that sustains not only the family but also the community as a whole—a shared endeavor that simultaneously remains deeply personal, reflecting the trajectories and life histories of individual homegarden tenders. Much like life itself, it renews with each passing year, in harmony with the natural rhythms of existence.

Introduction

The southern Andes, which are part of the Wallmapu, the ancestral territory of the Mapuche Indigenous people, are one of South America's most iconic territories in biocultural terms. In Chile, some of the most extensive remnants of the native forest of the central-southern part of the country are confined to mountainous areas (>600 meters altitude). These forests co-occur spatially with hundreds of Mapuche communities, along with the small-scale farms of many non-Indigenous families and, recently, a growing number of migrants of different cultures (Barreau et al. 2016; Zunino, Espinoza, and Vallejos-Romero 2016).

This territory embraces a unique intercultural setting in which to explore and learn about the importance of family agriculture for agrobiodiversity conservation and food sovereignty (Ibarra et al. 2020b). In these territories, family agriculture serves as a true refuge against a homogenizing wave of vast nonnative forest plantations, monoculture-intensive farming, introduced salmon farms, networks of highways, and large cities that has transformed a large part of central-southern Chile (Barreau et al. 2019). Here, homegardens are essential systems of family agriculture. These open-ended systems are the result of daily interactions among complex bodies of knowledge, practices, beliefs, and emotions associated with farming and food (Marchant Santiago et al. 2020). In addition, they take place in a landscape of reciprocal relationships between the people and active volcanoes, mountains, forests, rivers, springs, wetlands, and myriad animals, plants, fungi, and microorganisms that coinhabit it (Ibarra et al. 2022).

Homegardens are the backbone of family agriculture in the southern Andes. These small-scale productive systems can be home to great agrobiodiversity since they are multipurpose, where plants are grown not only for food, but also for medicinal, ornamental, and ritual purposes (Eyzaguirre and Linares 2010). They are also multistratified systems in which root and tuber crops, small annual and perennial plants, shrubs and small trees, and, in many cases, large trees all coexist (Galluzzi, Eyzaguirre, and Negri 2010). These homegardens, which are true extensions of the home, should be understood in a broader context as contributing to the heterogeneity of the Andean landscape. In general, scientific literature has paid great attention to the diversity of plants grown in homegardens in different countries, mainly in tropical areas (Eyzaguirre and Linares 2010; Kumar and Nair 2006; Norfolk, Eichhorn, and Gilbert 2013). Nonetheless, the information reported about biodiversity in homegardens in Chile remains scarce (Urra and Ibarra 2018). In Chile, plants of pre-Hispanic origin, many of them unique varieties adapted to local environmental conditions, are grown in homegardens, along with traditional medicines and foods (Ibarra et al. 2019).

Homegardens are constantly adapting to environmental changes (e.g., climate variability, water scarcity, the arrival of new species), to the historical context (e.g., new technologies, changes in the agricultural market, and state policies about agriculture and rural territories), and to the decisions of those who tend them. In homegardens it is possible, for example, to find cultivated plants that are considered to be "modern" or new to a territory alongside traditional varieties (landraces) and other native and exotic plants that may have been grown intentionally or not. The biodiversity found in homegardens is, in other words, a reflection of change, memory, and adaptation (Ibarra et al. 2021; Marchant Santiago et al. 2020; Nazarea 2006).

A homegarden is also a true open-air school, where plants and seeds, but also knowledge, memories, and learning, are exchanged (Barreau 2014; Celis 2003; Mellado 2014). To keep a homegarden is a skill that requires not merely experienced hands, but also vast knowledge about the ecology of the ecosystem (Celis 2003; Toledo 1994). Historically, this knowledge has been learned through experience and has been passed on orally through generations. Technological developments and mobility (virtual or real) over long distances, however, have diversified, added to, and hybridized local knowledge (Calvet-Mir et al. 2016). Knowledge is acquired no longer only from the mothers, grandmothers or neighbors who tend the homegarden, but also from the internet, state agricultural programs, NGOs, universities, books, workshops, and courses.

Gardening knowledge is also learned through creating networks of support and exchange with other gardeners and small-scale farmers of different origins and, therefore, with different ways of relating with the land. These networks, together with facilitating the exchange of agricultural knowledge among homegardeners from different cultures, could play a fundamental role in strengthening food sovereignty in intercultural contexts (Altieri and Toledo 2011; Wright 2014). Individuals' motivations for having a homegarden may differ, but their collective articulation could be critical in fostering informed, conscious, and empowered decision making about what, how, for whom, and how much to produce. This articulation can also favor biodiversity, care, and use of traditional varieties adapted to local conditions. Furthermore, it can support the family economy (e.g., cash income), access to healthy and high-quality food, and the genuine integration of diverse cultures in landscapes, such as the southern Andes, that are subject to constant social-environmental change (Caviedes et al. 2023; Marchant Santiago 2017).

In this chapter, we present the general results of a project based on three fundamental pillars of family agriculture in intercultural contexts: agrobiodiversity, learning, and food sovereignty. In particular, we (1) analyzed the general attributes and composition of plants in one hundred homegardens in the southern Andes of Chile; (2) explored sources of learning about gardening, management practices, and motivations for keeping a homegarden; and (3) facilitated a process of learning, exchange, and articulation between Mapuche and non-Mapuche campesino farmers with a long history in the territory and recent migrants to it.[1]

Methods

In the context of a larger project of social-environmental transdisciplinary research and place-based education in the southern Andes (Ibarra et al. 2022), between 2016 and 2020 we studied homegardens in the Andean zone of the La Araucanía region, including sectors of the municipal districts of Loncoche, Villarrica, Pucón, and Curarrehue. In this area, the Andean landscape is characterized by valleys whose floors are a heterogeneous mosaic of human settlements, farmlands, fragments of native forest, plantations of nonnative trees, shrubland, lakes, and rivers. As the land rises, native temperate forest increasingly dominates the landscape. The land of Mapuche and non-Mapuche campesino farmers is interspersed among large productive farms, public and private protected areas, and migrant and tourist country houses (Barreau 2014; Söhn 2012).

A sample of homegardens was selected using the snowball method (Newing 2011). It included fifty homegardens of Mapuche and non-Mapuche campesinos with a relatively long history in the area (>30 years living in the territory; *campesino homegardens*) and fifty homegardens of migrants to the area (*migrant homegardens*; figure 5.1). Non-Mapuche campesinos are understood as family nuclei of non-Indigenous people who were born, live, and work in the territory, in close contact with Mapuche families. Their agriculture and ways of living are derived from and incorporate the Mapuche agricultural system. Migrants are understood as families that represent a modern phenomenon of counter-urbanization, often referred to as "lifestyle migration." They moved from urban areas as adults, seeking places with unique natural and cultural attributes (Marchant Santiago 2017; Otero, Zunino, and Rodríguez 2017; Zunino, Espinoza, and Vallejos-Romero 2016; Zun-

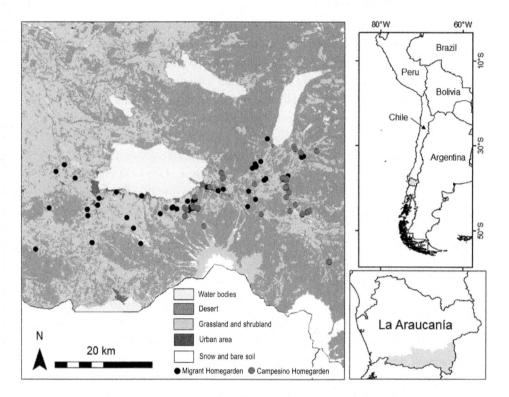

FIGURE 5.1 Location of the fifty campesino homegardens (gray dots) and the fifty migrant homegardens (black dots) who participated in the project. The box on the bottom right shows the location of the study area (in light gray), which includes Loncoche, Villarrica, Pucón, and Curarrehue municipal districts within the La Araucanía region, southern Andes of Chile.

ino and Hidalgo 2010). They are mostly professionals, sometimes of foreign origin, who have a paid job outside the agricultural sector. For the study, we considered homegardens from migrants (hereafter migrant homegardens). We specifically considered homegardens that were more than two years old, as this timeframe ensured that participants had gained sufficient gardening knowledge and experience.

A complete inventory of the plants grown intentionally was conducted in each homegarden through a guided tour with the gardener tending it. Information was also collected about the socioeconomic context and management practices of the homegarden through structured and semistructured interviews (see Bernard 2005; Newing 2011). To understand the main motivations for keeping a homegarden, weighted rankings were built with the following possible eight motivations: food, family economy, therapeutic/health, hobby, environmental education, conservation of agrobiodiversity, rescue of traditions/culture, and connection with nature (Newing 2011).

The one hundred gardeners were invited to participate in a program of free and volunteer workshops on agroecological principles and gardening. These were of two types: *technical* (led by an agroecologist) and *campesino-to-campesino* (led by an experienced local campesino). Two tours to Mapuche community agritourism initiatives also took place to strengthen the group's cohesion and share experiences.[2] Additionally, a *trafkintu*, or community gathering for exchanging seeds and plants as well as knowledge and experiences, was organized.

Results and Discussion

A great agrobiodiversity of plant species and varieties exists in the homegardens of the La Araucanía region, along with long-standing traditional agricultural practices. These homegardens are not static repositories; they are dynamic systems subject to a constant process of adaptation and innovation in intercultural contexts (Eyzaguirre and Linares 2010; Galluzzi, Eyzaguirre, and Negri 2010; Ibarra et al. 2021). Campesino homegardens have gradually incorporated "modern" varieties and innovative management practices, depending on the gardener's contemporary interests or new culinary tastes, or simply because they make the work easier or are consistent with the current context in terms of, for example, soil quality, the age of the gardener, or climate variability (Eyzaguirre and Linares 2010; Parraguez-Vergara et al. 2018). Migrant homegardens, in turn, reflect an interwoven diversity of agricultural and culinary cultures, depending on who tends them. These migrant homegardens, however, also adopt local practices, species, and varieties. The general characteristics of the two groups differ significantly (table 5.1), but in broad terms, the homegardens of the southern Andes are small-scale productive systems, surrounded by some protection (generally fences) and located near homes (figure 5.2). Our results support the notion that homegardens are an eminently feminine place (Celis 2003; Eyzaguirre and Linares 2010) since 89 percent of the homegardens in our sample were managed by women. Campesino homegardens are generally managed by women of advanced age and with long experience (see table 5.1).

TABLE 5.1 Attributes of fifty Mapuche and non-Mapuche campesino homegardens and fifty homegardens of migrants in the Araucanía region, southern Andes of Chile

Attribute[a]	Campesinos	Migrants
Age of gardener (years)	59 ± 13	49 ± 15
Experience of gardening (years)	35 ± 20	10 ± 10
Size of homegarden (m²)	394 ± 320	235 ± 227
Age of homegarden (years)	11 ± 12	5 ± 3
Contribution of homegarden to family consumption (scale 1–10)	9	6

[a]This table shows only attributes in which there were significant differences according to the Student's or Kruskal-Wallis t-test, depending on the type of data. The average ± standard deviation is shown for all attributes except "contribution to family consumption," for which the median is shown.

FIGURE 5.2 *Top*, examples of traditional Mapuche and non-Mapuche campesino homegardens. Photos by Antonia Barreau and Tomás Ibarra; *bottom*, homegardens of lifestyle migrants. Photos by Valentina Westermeyer and Lorena González.

Cultivated Plants: Homegardens as Dynamic Repositories of Biological Diversity

We found 284 species and 543 ethnovarieties of plants grown intentionally in the homegardens, with a great diversity of edible, aromatic, medicinal, and ornamental species.[3] The botanical families with the highest representation were Asteraceae ($n = 34$ species), Rosaceae ($n = 26$), Lamiaceae ($n = 23$), and Fabaceae ($n = 18$). The most common cultivated species, found in at least half of the homegardens, were cilantro, chives, oregano, peas, carrots, beets, fava beans, raspberries, chard, thyme, strawberries, parsley, tomatoes, common beans, potatoes, squash, and corn. The homegardens were diverse not only in species, but also in varieties, which are grown for their different colors, textures, flavors, and histories (Eyzaguirre and Linares 2010). For example, we recorded thirty-eight varieties of common beans (*Phaseolus vulgaris*) as well as more than six varieties of runner beans (*Phaseolus coccineus*; figure 5.3), twenty-five varieties of lettuces (*Lactuca sativa*), and more than twenty varieties of potatoes (*Solanum tuberosum*).

The most diverse campesino and migrant homegardens had a similar total of varieties: 100 and 107, respectively. Migrant homegardens, however, had a greater total diversity of species than the campesino homegardens (247 and 225 species, respectively, out of a total of 284 between the two groups). Similar results regarding a greater number of species in migrants' homegardens have been reported for other countries and reflect the flexibility of migrants who, when moving to a new place, take seeds of their favorite crops and incorporate locally grown varieties (Eyzaguirre and Linares 2010; Guerrero Peñuelas 2007; Neulinger, Vogl, and Alayón-Gamboa 2013). This greater plant diversity is not reflected in higher food production. Indeed, campesino gardeners reported a higher contribution to family consumption than migrants. In addition, half of the campesino gardeners produced surpluses for sale. This calls for attention since it has been reported that, in many cases, an increasing sale of produce from homegardens is detrimental to crop diversity (Eyzaguirre and Linares 2010; Howard 2006; Parraguez-Vergara et al. 2018).

Preliminary analysis indicates that older homegardens (range of 2–50 years) and larger ones for both groups contain a greater richness of plant species. In terms of plant composition, no clear patterns of similarity were found in the homegardens studied since those that were compositionally most similar had only 30 percent of their species in common. The great variability seen in floristic composition reflected the different emphasis that gar-

FIGURE 5.3 A sample of ethnovarieties of common beans (*Phaseolus vulgaris*) and runner beans (*Phaseolus coccineus*) recorded in homegardens in the southern Andes of Chile. Photo by Tomás Ibarra.

deners give to the use of plants for different purposes (Calvet-Mir et al. 2016; Eyzaguirre and Linares 2010; Galluzzi, Eyzaguirre, and Negri 2010). What is grown in each homegarden often reflects individual or family trajectories, rather than a specific sociocultural pattern. For example, one gardener relied on seeds inherited from her family and, in some cases, exchanged with neighbors, while another obtained their seeds during some visit to a relative in another place or through a workshop. Yet another could have received a plant as a thank-you from a friend or as an exchange in a *trafkintu*. We can, in other words, say that each homegarden and each plant bear their own stories; homegardens reflect different trajectories that will determine their composition, often without a common pattern between them.

Hands in the Earth: Learning Sources, Exchanges, and Management Practices

The transmission of knowledge is a dynamic and incessant process that depends on the circumstances of the home and changing ecological, social, and

TABLE 5.2 Sources of learning about the skill of gardening for campesinos and migrants in the southern Andes

Sources of learning	Campesinos (*n* = 50)	Migrants (*n* = 50)
Self-taught (%)	12	84
Training workshops (%)	2	24
Relatives (%)	90	28
Neighbors/local people (%)	0	52

Note: The participants could mention more than one source of learning; therefore, the sum of all percentages is higher than 100.

economic drivers (Howard 2006). The literature on traditional or campesino homegardens shows that plant knowledge is transmitted from an early age, first among women within a family nucleus and then among other close relatives (Celis 2003; Eyssartier, Ladio, and Lozada 2008; Howard 2006). In the southern Andes, campesinos identified female relatives (grandmothers, mothers, or older sisters) as their main source of knowledge, combined with some self-learning through experience (table 5.2). Migrants' sources of knowledge proved to be more diverse, which is related to factors such as not learning from an early age and not coming from a campesino family, greater mobility, and access to information and technologies as well as a lack of local learning networks. Many migrants defined themselves as self-taught through books, websites, and their own experimentation, but many also acknowledged seeking the support of a campesino neighbor or worker to tend their homegarden. This interaction between campesinos and migrants has enabled the latter to incorporate local varieties and the former's practices.

Trafkintus are also an opportunity to exchange knowledge and, at the same time, strengthen social networks that favor the conservation of agrobiodiversity (Nazarea 2005; Nazarea, Rhoades, and Andrews-Swann 2013; Peralta and Thomet 2013). When a seed or plant is exchanged, not only is plant material shared, but so is the associated knowledge, such as when to sow; what soil, water, and light the plant needs; and when to harvest (Calvet-Mir et al. 2016; Celis 2003; Peralta and Thomet 2013). As this is a long-standing activity, it is no coincidence that almost 50 percent of the campesinos participated in these seed exchanges. They also commonly exchange seeds with relatives and neighbors (Mellado 2014; Peralta and Thomet 2013). By contrast, only 24 percent of migrants mentioned participating in *trafkin-*

tus and, in many cases, were even unaware of their existence. They tended to buy seeds and seedlings and not to rely on obtaining seeds through exchange.

In terms of management practices, differences between the two groups to some extent reflect their origin, history, and socioeconomic situation (figure 5.4). Migrants' soil management and pest-control practices were more diverse. This can be attributed to their incorporation of practices learned from local campesinos and techniques derived from other sources of learning, most of them agroecological. Differences were also observed in the use of agrochemicals, particularly fertilizers and pesticides, of which the campesinos made greater use. It is important to note that 72 percent of the campesinos interviewed (compared to only 12 percent of the migrants) were beneficiaries of state agricultural subsidy programs, which generally provided free agrochemicals (Candia 2013; Clark 2011; Parraguez-Vergara et al. 2018). In some cases, we observed agrochemicals stacked up in the storehouses of gardeners who opt for "clean or organic" production.

Other campesino gardeners who received agrochemicals from state agricultural subsidy programs said they sold them or gave them away. This indicates that these subsidies, albeit providing opportunities, are often not aligned with organic production or the beneficiaries' intentions. As one

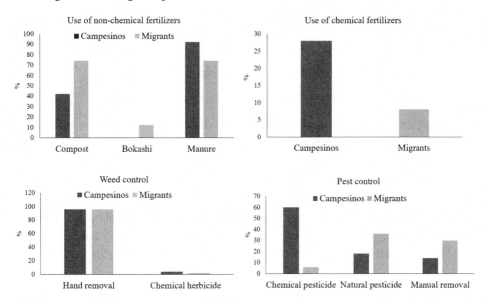

FIGURE 5.4 Management practices in campesino homegardens (*n* = 50) and migrant homegardens (*n* = 50) in the southern Andes of Chile.

campesino observed, "State agricultural subsidy programs try to be a help for our soils; however, they end up ruining them." This disconnect between state subsidy programs and small family farming has been reported in the literature as a possible threat to biocultural diversity when local knowledge and practices are not valued or promoted (Clark 2011; Jacobi et al. 2017; Parraguez-Vergara et al. 2018). In addition, many campesino gardeners have had fewer opportunities for formal studies and less access to information about the potential negative effects of these agrochemicals. Among campesinos, certain traditional practices, such as the use of manure, adherence to lunar cycles, secret ways of avoiding frost, and the cultivation of certain magical or oracular plants, were also more common. The migrants, on the other hand, mostly adhered to the agroecological movement and its principles (Marchant Santiago 2017; Otero, Zunino, and Rodríguez 2017; Zunino, Espinoza, and Vallejos-Romero 2016).

In economic terms, migrants reported buying more inputs such as seeds, seedlings, and manure than campesinos, who, as part of a broader agroforestry system, are more self-sufficient (e.g., farmers generate fertilizer from their own animals) and tend not to buy inputs for their homegardens (Engels 2002; Eyzaguirre and Linares 2010). In addition, many campesinos generate surplus seeds, seedlings, medicines, food, and manure, which they then exchange or sell.

Cultivating Sovereignty: Motivations for Gardening and the Articulation of Local Networks

Food sovereignty is exercised for very personal and collective motives related to the right to produce food that is ecologically, socially, and economically appropriate to the circumstances and contexts (Pimbert 2018). This study found that food production for family consumption was the main motivation for keeping a homegarden among both campesinos and migrants (figure 5.5). For the campesinos, unlike migrants, however, the homegarden's contribution to the family economy (cash income) was equally important. This is interesting, given the large number of ornamental plants, without commercial value, they cultivate in their homegardens. For migrants, the second and third most important motivations were the homegarden as a therapeutic place associated with health and as a means of (re)connecting with nature. This is consistent with results found for gardeners in the Catalan

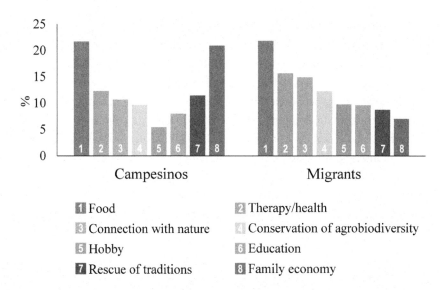

FIGURE 5.5 Results of weighted ranking of the motivations of Mapuche and non-Mapuche campesinos and of migrants for keeping a homegarden in the southern Andes of Chile.

Pyrenees, a rural population in an industrialized country, whose motivations were related more to the positive effects for well-being than to economic considerations (Calvet-Mir et al. 2016; Reyes-García et al. 2012). This also reflects the more personal aspirations of lifestyle migrants to this mountain area in southern Chile (Marchant Santiago 2017; Otero, Zunino, and Rodríguez 2017; Zunino and Hidalgo 2010).

Through the practices of campesinos and migrants, food sovereignty can permeate the territory beyond their specific individual motivations. Campesinos are often motivated by their interest in producing food and medicinal plants to give away or sell in their communities or at the local market. As homegardens are a living repository of species and varieties that, year after year, are reproduced, stored, and exchanged, they also serve to exercise and strengthen seed sovereignty (Nazarea 2005; Nazarea, Rhoades, and Andrews-Swann 2013; Peralta and Thomet 2013). Many lifestyle migrants aspire to learn more about the Mapuche culture and strengthen the local agricultural culture (Marchant Santiago 2017), and some become involved in local government activities, rural schools, and community organizations to promote intercultural education (Ibarra et al. 2020a; Zunino, Espinoza, and

Vallejos-Romero 2016). It is, therefore, vital to strengthen the social fabric formed by gardeners of different origins who currently inhabit the territory. To strengthen this network, we organized workshops and two tours to Mapuche community agritourism initiatives to enable the gardeners to meet and learn about agroecological principles and associated initiatives (e.g., associative tourism, beekeeping, and local/traditional gastronomy).

In both the technical and campesino-to-campesino workshops, experiences of integrated soil and pest management were exchanged and biopreparations were made, according to agroecological principles (figure 5.6). In addition, a *trafkintu* was organized, to which the hundred gardeners involved in this study were invited to exchange seeds, plant propagules, knowledge, and experiences as well as strengthen their support networks (see figure 5.6).

FIGURE 5.6 *Top, trafkintu* organized as part of the project in which Mapuche and non-Mapuche campesinos and migrants exchanged seeds, plants, and experiences. Photos by Cristóbal Saavedra; *bottom left,* gardeners participating in a biopreparations workshop, led by Lorena González. Photo by Antonia Barreau; *bottom right,* gardeners discussing the structure of the soil and care for it in the workshop on agroecological management led by Heraldo Carvacho. Photo by Julián Caviedes.

This incipient articulation has generated new trust and friendships, and it is hoped that, with ongoing long-term work, it will foster social cohesion, appreciation, and respect for the Mapuche and non-Mapuche campesino territory. This fabric can also be a true seed of hope for the creation of a network of exchange (sale or barter) that favors the local circular economy and the flow of local varieties that need to be widely propagated and consumed to ensure their conservation (Galluzzi, Eyzaguirre, and Negri 2010).

Conclusions

Homegardens in the southern Andes, or Wallmapu, the ancestral land of the Mapuche people, contain an extraordinary agrobiodiversity of plants of which there was previously no record in southern Chile (Urra and Ibarra 2018). These small agricultural systems can be true individual refuges for the revitalization of biocultural memory (Barreau and Ibarra 2019; Marchant Santiago et al. 2020; Nazarea 2006). Beyond this valuable individual role, the collective articulation of these homegardens and those who cultivate them could become a powerful network for strengthening family agriculture and maintaining local traditional varieties and practices. Even more importantly, it can strengthen food sovereignty and serve as an engine of intercultural respect and integration.

While lifestyle migrants can be a source of agrobiodiversity and agricultural innovation, Mapuche and non-Mapuche campesinos are a source of local expert knowledge and traditional varieties. Cooperation between these two types of gardeners and their integration can be harnessed to the conservation of local agrobiodiversity and its intercultural manifestations. This integration may also be conducive to different modes of transmission of agricultural knowledge and practices, fostering diversity and the resilience of a territory that is experiencing important demographic and social-environmental changes.

Acknowledgments

We are grateful to Romina Urra, Natalia Pessa, Fernanda Barreau, Valentina Undurraga, Daniela Westermeyer, and Tomás Altamirano for their support in the field. Francisca Santana prepared the cartography (figure 5.1), and Lucía Ferreira provided invaluable support in the project's management. We are indebted to the people who facilitated the workshops and tours: Ana Ayelef, Patricia Ayelef, Juan Caniucura, Stephanie Carmody, Heraldo Carvacho, Angélica Chincolef, Rosa Huaiquifil, Lorena González, Jerry Laker, lonko don Juan Huilipan, and Manuel Maribur (Mapuche Trekan). We would particularly like to thank the gardeners who shared their time, knowledge, and experience during visits to their homegardens and in the project's workshops and *trafkintu*. We thank the support from Fundación para la Innovación Agraria through the project *Huerta andina de La Araucanía como patrimonio biocultural: Un enfoque agroecológico y agroturístico* (PYT-2016–0347); the Center for Intercultural and Indigenous Research (ANID/FONDAP/15110006); the Center of Applied Ecology and Sustainability (ANID PIA/BASAL FB0002), ANID/REDES (190033); the Cape Horn International Center for Global Change Studies and Biocultural Conservation (ANID PIA/BASAL PFB210018); and ANID/Fondecyt Regular 1200291. We also thank the organizations associated with the project: Comunidad Indígena Rayen Lelfun, Mapuche; Red de Agroemprendedores de Pichares; Aldea Lacustre; Grupo Guías Cañe; and Kodkod: Lugar de Encuentros.

Resumen

En el sur de los Andes conviven huertas de campesinos mapuche y no-mapuche, y las de un número creciente de inmigrantes. Exploramos la agrobiodiversidad, fuentes de aprendizaje, prácticas de manejo y soberanía en 100 huertas familiares (50 campesinas y 50 migrantes) en La Araucanía andina del sur de Chile. Utilizando metodologías mixtas, encontramos una extraordinaria diversidad de plantas (284 especies y 543 etnovariedades). Las huertas de migrantes presentaron una mayor diversidad de plantas. Para los campesinos, las fuentes de aprendizaje fueron principalmente familiares, mientras que para migrantes fueron más diversas dada su mayor movilidad, acceso a información y tecnologías. Para ambos grupos, la principal motivación para cultivar una huerta fue el aporte alimentario, pero, para campesinos, también lo fue el aporte económico. Mientras los migrantes son fuente de agrobiodiversidad e innovación, los campesinos son fuente de conocimiento local experto y variedades tradicionales. Proponemos e implementamos acciones para la integración entre agricultores para favorecer la agrobiodiversidad y la soberanía alimentaria en contextos interculturales del sur de Sudamérica.

Notes

1. Although the literature differentiates the terms *Indigenous* and *campesino* (with the latter generally implying non-Indigenous), we refer here to Mapuche and non-Mapuche campesinos since the Mapuche farmers who inhabit rural areas of the southern Andes of Chile self-identify as campesinos.

2. See Mapuche Trekan, accessed March 17, 2024, https://www.mapuche-trekan .com/.

3. The varieties correspond to what the literature defines as *ethnovarieties* since they are identified as such by the homegardeners themselves. Therefore, two or more ethnovarieties could be the same variety for one species.

References

Altieri, Miguel A., and Victor Manuel Toledo. 2011. "The Agroecological Revolution in Latin America: Rescuing Nature, Ensuring Food Sovereignty and Empowering Peasants." *Journal of Peasant Studies* 38 (3): 587–612. https://doi.org/10.1080/030 66150.2011.582947.

Barreau, Antonia. 2014. "Narrating Changing Foodways: Wild Edible Plant Knowledge and Traditional Food Systems in Mapuche Lands of the Andean Temperate Forests, Chile." PhD diss., University of British Columbia, Vancouver.

Barreau, Antonia, and María Ignacia Ibarra. 2019. "Mujeres Mapuche y Huertas Andinas: Espacios de Fertilidad, Soberanía y Transmisión de Saberes." In *Huertas Familaires y Comunitarias: Cultivando Soberanía Alimentaria*, edited by José Tomás Ibarra, Julián Caviedes, Antonia Barreau, and Natalia Pessa, 126–37. Santiago, Chile: Ediciones UC.

Barreau, Antonia, José Tomás Ibarra, Felice S. Wyndham, and Robert A. Kozak. 2019. "Shifts in Mapuche Food Systems in Southern Andean Forest Landscapes: Historical Processes and Current Trends of Biocultural Homogenization." *Mountain Research and Development* 39 (1): 1–18. https://doi.org/10.1659/MRD-JOURNAL -D-18-00015.1.

Barreau, Antonia, José Tomás Ibarra, Felice S. Wyndham, Alejandro Rojas, and Robert A. Kozak. 2016. "How Can We Teach Our Children if We Cannot Access the Forest? Generational Change in Mapuche Knowledge of Wild Edible Plants in Andean Temperate Ecosystems of Chile." *Journal of Ethnobiology* 36 (2): 412–32. https://doi.org/10.2993/0278-0771-36.2.412.

Bernard, H. Russell. 2005. *Research Methods in Anthropology: Qualitative and Quantitative Approaches*. 4th ed. Oxford: Altamira.

Calvet-Mir, Laura, Carles Riu-Bosoms, Marc González-Puente, Isabel Ruiz-Mallén, Victoria Reyes-García, and José Luis Molina. 2016. "The Transmission of Home Garden Knowledge: Safeguarding Biocultural Diversity and Enhancing Social-Ecological Resilience." *Society and Natural Resources* 29 (5): 556–71. https://doi .org/10.1080/08941920.2015.1094711.

Candia, Javier. 2013. "Campesinos Advierten Situación Crítica Por Uso de Agrotóxicos." *DiarioUChile*, December 3, 2013.

Caviedes, Julián, José Tomás Ibarra, Laura Calvet-Mir, and André Braga Junqueira. 2023. "'Listen to Us': Small-Scale Farmers' Understandings of Social-Ecological Changes and Their Drivers in Important Agricultural Heritage Systems." *Regional Environmental Change* 23, art. 158. https://doi.org/https://doi.org/10.1007/s10113-023-02145-9.

Celis Salamero, María Angélica. 2003. "Conversaciones Con El Territorio Desde La Interculturalidad. Las Huertas Femeninas Como Espacio de Conversación." Master's thesis, Universidad de la Frontera, Temuco, Chile.

Clark, Timothy D. 2011. "Putting the Market in Its Place: Food Security in Three Mapuche Communities in Southern Chile." *Latin American Research Review* 46 (2): 154–79.

Engels, Jan. 2002. "Home Gardens—a Genetic Resources Perspective." In *Proceedings of the Second International Home Gardens Workshop: Contribution of Home Gardens to in Situ Conservation of Plant Genetic Resources in Farming Systems, 17–19 July 2001*, edited by Jessica Watson and Pablo B. Eyzaguirre, 193. Rome: International Plant Genetic Resources Institute.

Eyssartier, Cecilia, Ana H. Ladio, and Mariana Lozada. 2008. "Cultural Transmission of Traditional Knowledge in Two Populations of North-Western Patagonia." *Journal of Ethnobiology and Ethnomedicine* 4, art. 25. https://doi.org/10.1186/1746-4269-4-25.

Eyzaguirre, Pablo B., and Olga F. Linares. 2010. *Home Gardens and Agrobiodiversity*. 2nd ed. Washington, D.C.: Smithsonian Institution Press.

Galluzzi, Gea, Pablo Eyzaguirre, and Valeria Negri. 2010. "Home Gardens: Neglected Hotspots of Agro-Biodiversity and Cultural Diversity." *Biodiversity and Conservation* 19 (13): 3635–54. https://doi.org/10.1007/s10531-010-9919-5.

Guerrero Peñuelas, Adriana G. 2007. "El impacto de la migración en el manejo de solares campesinos, caso de estudio La Purísima Concepción Mayorazgo, San Felipe Del Progreso, Estado de México." *Investigaciones Geográficas* 63:105–24.

Howard, Patricia L. 2006. "Gender and Social Dynamics in Swidden and Home-gardens in Latin America." In *Tropical Homegardens: A Time-Tested Example of Sustainable Agroforestry*, edited by B. M. Kumar and P. K. R. Nair, 159–82. Dordrecht: Springer. https://doi.org/10.1007/978-1-4020-4948-4.

Ibarra, José Tomás, Antonia Barreau, Julián Caviedes, Natalia Pessa, and Romina Urra. 2019. "Huertas Familiares Tradicionales y Emergentes: Cultivando Biodiversidad, Aprendizaje y Soberanía Desde La Interculturalidad." In *Huertas Familiares y Comunitarias: Cultivando Soberanía Alimentaria*, edited by José Tomás Ibarra, Julián Caviedes, Antonia Barreau, and Natalia Pessa, 224. Santiago, Chile: Ediciones UC.

Ibarra, José Tomás, Antonia Barreau, Julián Caviedes, Natalia Pessa, Jeannette Valenzuela, Sylvia Navarro-Manquilef, Constanza Monterrubio-Solís, Andrés Ried, and José Cristóbal Pizarro. 2020a. "Listening to Elders: Birds and Forests as Intergenerational Links for Nurturing Biocultural Memory in the Southern Andes." In *Latin American Transnational Children and Youth: Experiences of Nature and Place, Culture and Care Across the Americas*, edited by Victoria Derr and Yolanda Corona, chapter 13. Abingdon: Routledge.

Ibarra, José Tomás, Antonia Barreau, Carla Marchant, Jorge A. González, M. Oliva, M. E. Donoso-Correa, B. Antaki, Constanza Monterrubio-Solís, and Fausto O. Sarmiento. 2020b. "Montology: An Integrative Understanding of Mountain Foodscapes for Strengthening Food Sovereignty in the Andes." In *The Elgar Companion to Geography, Transdisciplinarity and Sustainability*, edited by F. O. Sarmiento and L. Frolich, 391–405. Cheltenham: Edward Elgar Publishing.

Ibarra, José Tomás, Julián Caviedes, Tomás A. Altamirano, Romina Urra, Antonia Barreau, and Francisca Santana. 2021. "Social-Ecological Filters Drive the Functional Diversity of Beetles in Homegardens of Campesinos and Migrants in the Southern Andes." *Scientific Reports* 11, art. 12462. https://doi.org/10.1038/s41598-021-91185-4.

Ibarra, José Tomás, Robert Petitpas, Antonia Barreau, Julián Caviedes, Josefina Cortés, Gabriel Orrego, Gonzalo Salazar, and Tomás A. Altamirano. 2022. "Becoming Tree, Becoming Memory: Social-Ecological Fabrics in Pewen (*Araucaria araucaria*) Landscapes of the Southern Andes." In *The Cultural Value of Trees: Folk Value and Biocultural Conservation*, edited by Jeffrey Wall, 15–31. Abingdon: Routledge.

Jacobi, Johanna, Sarah-Lan Mathez-Stiefel, Helen Gambon, Stephan Rist, and Miguel Altieri. 2017. "Whose Knowledge, Whose Development? Use and Role of Local and External Knowledge in Agroforestry Projects in Bolivia." *Environmental Management* 59 (3): 464–76. https://doi.org/10.1007/s00267-016-0805-0.

Kumar, B. M., and P. K. R. Nair, eds. 2006. *Tropical Homegardens: A Time-Tested Example of Sustainable Agroforestry*. Vol. 3 of *Advances in Agroforestry*. Dordrecht: Springer Netherlands.

Marchant Santiago, Carla. 2017. "Lifestyle Migration and the Nascent Agroecological Movement in the Andean Araucanía, Chile: Is It Promoting Sustainable Local Development?" *Mountain Research and Development* 37 (4): 406–14. https://doi.org/10.1659/MRD-JOURNAL-D-17-00036.1.

Marchant Santiago, Carla, Nicolás Fuentes Acuña, Santiago Kaulen Luks, and José Tomás Ibarra. 2020. "Local Knowledge in Montane Homegardens in the Southern Andes: A Refuge of Mapuche Pewenche Biocultural Memory." *Pirineos* 175:e060. https://doi.org/10.3989/PIRINEOS.2020.175010.

Mellado, M. A. 2014. "¡Eran Raíces! Relaciones Sociales En Las Huertas Familiares Mapuche Del Lago Neltume, Panguipulli." Thesis, Universidad Austral de Chile, Valdivia.

Nazarea, Virginia D. 2005. *Cultural Memory and Biodiversity*. Tucson: University of Arizona Press.

Nazarea, Virginia D. 2006. "Local Knowledge and Memory in Biodiversity Conservation." *Annual Review of Anthropology* 35 (1): 317–35. https://doi.org/10.1146/annurev.anthro.35.081705.123252.

Nazarea, Virginia D., Robert E. Rhoades, and Jenna E. Andrews-Swann, eds. 2013. *Seeds of Resistance, Seeds of Hope: Place and Agency in the Conservation of Biodiversity.* Tucson: University of Arizona Press.

Neulinger, Korinna, Christian R. Vogl, and José A. Alayón-Gamboa. 2013. "Plant Species and Their Uses in Homegardens of Migrant Maya and Mestizo Smallholder Farmers in Calakmul, Campeche, Mexico." *Journal of Ethnobiology* 33 (1): 105–24. https://doi.org/10.2993/0278-0771-33.1.105.

Newing, Helen. 2011. *Conducting Research in Conservation: A Social Science Perspective.* New York: Routledge.

Norfolk, Olivia, Markus P. Eichhorn, and Francis Gilbert. 2013. "Traditional Agricultural Gardens Conserve Wild Plants and Functional Richness in Arid South Sinai." *Basic and Applied Ecology* 14 (8): 659–69. https://doi.org/10.1016/j.baae.2013.10.004.

Otero, Adriana María, Hugo Marcelo Zunino, and Mariana Rodríguez. 2017. "Las tecnologías socioculturales en los procesos de innovación de los migrantes de amenidad y por estilos de vida: El caso del destino turístico de Pucón, Chile." *Revista de Geografía Norte Grande* 233 (67): 211–33. https://doi.org/10.4067/S0718-34022017000200011.

Parraguez-Vergara, Elvis, Beatriz Contreras, Neidy Clavijo, Vivian Villegas, Nelly Paucar, and Francisco Ther. 2018. "Does Indigenous and Campesino Traditional Agriculture Have Anything to Contribute to Food Sovereignty in Latin America? Evidence from Chile." *International Journal of Agricultural Sustainability* 16 (4–5): 326–41. https://doi.org/10.1080/14735903.2018.1489361.

Peralta, Cristián, and Max Thomet, eds. 2013. *Curadoras de semillas: El arte de conservar semillas.* Temuco, Chile: Ediciones CETSUR.

Pimbert, Michel P. 2018. *Food Sovereignty, Agroecology and Biocultural Diversity: Constructing and Contesting Knowledge.* New York: Routledge.

Reyes-García, Victoria, Laura Aceituno, Sara Vila, Laura Calvet-Mir, Teresa Garnatje, Alexandra Jesch, Juan José Lastra, Montserrat Parada, Montserrat Rigat, Joan Vallès, and Manuel Pardo-De-Santayana. 2012. "Home Gardens in Three Mountain Regions of the Iberian Peninsula: Description, Motivation for Gardening and Gross Financial Benefits." *Journal of Sustainable Agriculture* 36 (2): 249–70.

Söhn, Lisa. 2012. "Landowner Attitudes Towards a Chicken-Killing Neotropical Felid in the Araucanía Region of Southern Chile." Master's thesis, Technische Universität München, Munich, Germany.

Toledo, Victor M. 1994. "La apropiación campesina de la naturaleza: Un análisis etnoecológico." PhD diss., Universidad Nacional Autónoma de México.

Urra, Romina, and José Tomás Ibarra. 2018. "Estado del conocimiento sobre huertas familiares en Chile: Agrobiodiversidad y cultura en un mismo espacio." *Etnobiología* 16 (1): 31–46.

Wright, Sarah. 2014. "Food Sovereignty in Practice: A Study of Farmer-Led Sustainable Agriculture in the Philippines." In *Globalization and Food Sovereignty: Global* and *Local Change in the New Politics of Food*, edited by P. Andrée, J. Ayres, M. J. Bosia, and M. J. Massicotte, 199–227. Toronto, ON: University of Toronto Press.

Zunino, Hugo Marcelo, Lorena Espinoza Arévalo, and Arturo Vallejos-Romero. 2016. "Los migrantes por estilo de vida como agentes de transformación en la Norpatagonia chilena." *Revista de Estudios Sociales* 1 (55): 163–76. https://doi.org/10.7440/res55.2016.11.

Zunino, Hugo Marcelo, and Rodrigo Hidalgo. 2010. "En busca de la utopía verde: Migrantes de amenidad en la Comuna de Pucón, IX región de la Araucanía, Chile." *Scrip Nova* 14 (331): 1–14.

CHAPTER SIX

Balancing Legitimated Livelihoods with Embodied Craft

Practices of U.S. Immigrant Farmers

EMILY RAMSEY

It goes without saying that farming is a tough way of life. The bar is extremely high to start farms for those who already own some means of production, and even higher for those who lack land, capital, and equipment and start out largely from scratch. As noted by Sebastian, an Ecuadorean immigrant farmer living in Florida, farming can seem "like a system of getting into debt." This is especially true when one jumps on the technological treadmill and turns to "selling and having to make up the cost of these crazy machines that help you [farm]." Statements like his point not only to farming's obstacles, but to its market-oriented demands.

For U.S.-based immigrant farmers who face unique economic and logistic challenges in beginning and maintaining farming operations, agriculture's market aspects can easily seem to dominate. More often than not, these market-based aspects—and the myriad steps beginning farmers must take so they can start to engage with consumers or wholesalers in markets— require varying degrees of legibility. Farmers are not simply persons good at tending animals or sticking seeds in the ground and making plants grow. Rather, as Deborah Fitzgerald (2003) notes, for almost a century agriculture has been permeated by an industrial logic dictating that farmers should also be businesspersons whose activities are driven by efficiency, rationality, and the goal of maximizing yield. This state of efficiency is achieved, in part, by making the farm and its activities increasingly legible through the collection of data for control and manipulation, legibility that permits them to access loans, services, subsidies, and programs. Such legibility parallels, in certain ways, the systematic collection, evaluation, and storage of germplasm in biodiversity conservation. That is, as Nazarea (1998) argues, although gene banks perform the critical task of preserving plant species and varieties from around the world, they do so in a way that reduces plant data to a basic catalog of easily searched, compared, and controlled elements, ignoring valuable in-depth cultural knowledge of those most familiar with the plants.

Despite desires for farming efficiency, it is equally well recognized that farming is also a craft—a sensuous and often deeply embodied activity, involving meaningful choices that connect farmers to plants, places, memories, and people (Gerodetti and Foster 2016; Graham and Connell 2006; Mazumdar and Mazumdar 2012). This volume examines those often-marginal places where people draw on meaning, memory, and enchantment to hold onto the plants and practices that connect them to other people, places, and times. These acts are what constitute the embodiment of biodiversity. Here, I engage in this conversation by focusing on Latinx immigrant farmers who contribute to not only conservation but the relational work of embodying biodiversity, even as they navigate the competing legibility demands of the contemporary U.S. agrifood system.

Embodiment, rooted frequently in intangible elements like meaning and memory, and inscription, the process of recording details through tangible and concrete descriptors, often seem at odds. Recognizing this, in this chapter I examine the ways in which Latinx immigrant farmers walk a line between engaging in meaningful, embodied practice and seeking the legitimization of "the Market" that often reflects an aura of sovereignty. I critique seemingly monolithic portrayals of a singular Market (designated by a capital *M*), suggesting that the distinct and multivalent markets (plural and designated by a lowercase *m*) reflect a "diverse economy" (St. Martin, Roelvink, and Gibson-Graham 2015, 4) and the reality Latinx immigrant farmers encounter and navigate. Through several ethnographic vignettes, I illustrate how Latinx immigrant farmers weave together threads of meaning in sometimes creative ways. As they do so, they engage in activities that inscribe them within markets but at the same time allow them the latitude to farm in ways that are personally meaningful and embodied. Such modes of farming connect them to landscapes, plants, and memories from their homes while also emplacing them in the United States.

Thus, following Schmid (2020), I contribute to the growing body of literature that challenges the common conflation of Latinx people in agriculture with farmworkers by instead focusing on their creative farming operations and practices. Moreover, following DeGuzmán's (2019) call to decolonize botany by centering Latinx botanical epistemologies, I hope to shed light on the contributions these farmers are making not only to their local foodsheds but also to preserving and extending plant varieties and knowledge about them. Ultimately, I aim to show that, as transnational brokers of sometimes unfamiliar crops and flavors, many Latinx immigrant farmers are working hard to connect their livelihood to sensuously embodied craft and appear to do so with success. These actions and successes are critical as they push back against dictates and demands for market legibility, often touted as the solution to Latinx farmers' struggles, complicating our understanding of the worlds they—and many other small farmers—navigate.

This chapter draws on six weeks of preliminary research carried out in the summer of 2018 in New York, Florida, and Louisiana with commercially oriented immigrant farmers. It also draws on an additional month of field-work in early 2020 in Florida in the early days of the COVID-19 pandemic. Each of the farmers interviewed and mentioned herein immigrated to the United States during their lifetimes from Mexico, the Caribbean, or Central or South America. These stages of research ultimately included conversations and interviews with twenty-five farmers and eight organization directors and outreach specialists at five organizations that connect immigrant farmers with land, resources, markets, and training. The conversations and ideas here, however, represent a smaller subset of those interactions.

Latinx Farmers in the United States

Over the last several decades, the average age of U.S.-based small farmers has steadily increased, raising concerns about who will take up the mantle of future agricultural production (Burns and Kuhns 2016). A noteworthy population increasingly meeting this call are Latinx immigrants. According to data extrapolated from the USDA Census of Agriculture in 2012 and 2017, numbers of Latinx farm operators (those who claim Latin American heritage) have increased 36 percent between 2007 and 2017 to now over 112,000 (U.S. Department of Agriculture 2014, 2019). These statistics, however, significantly underrepresent the actual numbers of Latinx farmers, due to factors that prevent them from registering, like language and literacy barriers, immigration status concerns, apprehension about enrolling in a government agency, and informal farming arrangements that promote their invisibility (Garcia, Lopez-Ariza, and Marinez 2008; Minkoff-Zern 2019; Minkoff-Zern and Sloat 2017).

Latinx immigrant farmers beginning or managing their own farms is not a new phenomenon. Rather, it is one whose size and visibility has steadily risen in the last several decades. Not surprisingly, these farms are most numerous in regions such as the American Southwest, or in areas with a historical migration of persons from Mexico, Central and South America, or the Caribbean (U.S. Department of Agriculture National Agricultural Statistical Service 2019). Especially along the West Coast, many of these farmers were formerly farmworkers or farm managers who became proprietors when the farmers they worked for retired or exited farming (Korsunsky 2020; Minkoff-Zern 2019). Other immigrant farmers have pursued agricultural enterprises like nurseries, growing plant starts, or cultivating fresh fruits and vegetables, primarily because initial capital outlays are lower, and permanent alterations to rented plots of land are unnecessary (Minkoff-Zern 2019).

In past decades, few studies recognized the phenomenon of Latinx immigrant farmers (Wells [1990, 1996] being one notable exception). Recently, however, academics and practitioners have increasingly turned their attention to this phenomenon (see Calo 2018; Korsunsky 2020; Minkoff-Zern 2017, 2018, 2019; Minkoff-Zern and Sloat 2017; Schmid 2019, 2020). The handful of studies in the last fifteen years that have focused on Latinx farm owners or operators have tended to examine (1) the reasons for their relative invisibility in the agricultural landscape (Garcia and Marinez 2005); (2) the obstacles they face (Swisher, Brennan, and Shah 2007; Vásquez-León 2009); and (3) the culturally based social networks they construct to mitigate risk and overcome obstacles (Flora et al. 2011; Vásquez-León 2009). These foci are not surprising, as myriad obstacles facing immigrant farmers are well attested. The bar for these farmers to start or maintain their own farms is high. Farmers need access to capital, equipment, land, and markets—or what one director of a New York–based organization supporting aspiring immigrant farmers I interviewed called "the famous 'triple-T:' *tierra, tractores, y tiendas* [land, tractors, and markets]." For organizations assisting Latinx immigrant farmers, aid often takes the form of education on topics ranging from agricultural sustainability to best business practices, attempting to address a lack of knowledge from which these farmers are assumed to suffer (Calo and De Master 2016; Hightower, Niewolny, and Brennan 2013; Niewolny and Lillard 2010).

Within the most recent social science scholarship focused on Latinx immigrant farmers, there is a notable shift. These studies acknowledge the challenges immigrant farmers face while attempting to situate these farmers within the larger U.S. agrifood complex. Their biggest contribution, however, touches on not only why the bar for entry is so high but begins to document the similarities and diversity among Latinx farmers across the United States to understand the differential impact of structural inequality on them. For example, scholars like Adam Calo (2018) argue that outreach focused on farm management education erroneously assumes that farmers inherently lack knowledge when, in fact, structural issues, like access to land, are far bigger challenges for them. In Laura Anne Minkoff-Zern's recent book *The New American Farmer*, she goes a step further, focusing on the impact racialization and the notion of "Mexican exceptionalism" has on Latinx immigrant farmworkers turned farmers (2019, 31). Their racialization places severe structural limitations on them, limiting their ability to benefit from USDA programs supporting small farmers or to obtain organic certification. Despite this, she argues, their agricultural background prior to immigration allows them to farm in inherently more sustainable ways, making them well suited to current alternative food movements' goals (see also Minkoff-Zern, Walsh, and Ludden 2019).

As noted above, however, Latinx farmers are a diverse group. Consequently, Alex Korsunsky draws on his work among Oregon farmers to challenge the homogeneous picture Minkoff-Zern paints, arguing that to see all as "inclined by virtue of their agrarian heritage" toward sustainable agricultural practices misses the significant number of Latinx immigrant farmers engaged in conventional agriculture (2020, 116). To focus solely on Latinx farmers who engage in sustainable farming practices may support exemplars that food sovereignty advocates seek to champion, but Korsunsky also fears that it falls into anthropology's romantic tendency to exoticize the subaltern as traditional and utopian. These recent contributions inform this chapter in various ways. Here, I too am contributing to the diverse portrait of Latinx immigrant farmers across the United States by adding voices largely from the Southeast. Likewise, I am deeply interested in the sustainable practices many of those I met were performing. Korsunsky's caution not to excessively romanticize these activities, however, offers a valuable reminder to set Latinx farmers in the context of a larger, inherently complex set of choices they must make and circumstances they must navigate.

The Value and Limitations of a
Market-Deterministic Perspective

As noted above, scholarship on Latinx immigrant farmers to date has focused on the challenges and obstacles facing Latinx immigrant farmers,
especially the structural inequality derived from their positionality as immigrants at the bottom of the bottom of the U.S. agrifood chain. Much of
this attention also rightly emphasizes U.S. agricultural markets' dominance
and seemingly deterministic impact on Latinx immigrant farmers' success.
Building on this, a focus on markets can importantly reflect unique and
creative forms of market-oriented knowledge Latinx farmers develop. Mary
Elizabeth Schmid's work among Latinx farming families in Southern Appalachia is one such example. She describes what she labels a "*commercio* mētis"
among the Latina women she observed creatively and profitably managing
the financial aspects of their families' farms (2020, 162). By this term, she
refers to experientially derived "practical commerce skills and wisdom" that
help them assign produce quality grades and then move vast quantities of
perishable fresh fruits and vegetables to buyers, in turn ensuring that these
Latina women are paid fairly (162).

Still, the onus lies on immigrant farmers to be equal parts growers and
businesspeople. As noted at the outset of this chapter, this reflects a century-
long project in the United States (Fitzgerald 2003), propelled by an incisive
rationality and the farm data that drive farmers' decisions. As agriculture
began to mature under a capitalist framework, it became necessary to jump
on the "technological treadmill" (Cochrane 1993; Rissing 2016), generating
technologies that have allowed farmers to control and overcome natural agricultural cycles (Boyd and Watts 1997; Lewontin 2000). Simultaneously, a
larger farm size became essential for most farmers to compete and succeed
(Fitzgerald 2003), famously summarized by U.S. Secretary of Agriculture
Earl Butz's often-cited 1973 statement to "get big or get out." These premises encapsulate what James C. Scott calls the "catechism of high-modernist
agriculture," a paradigm dominant in the last half of the twentieth century
(1998, 270). This paradigm is characterized by monoculture, mechanization,
standardization, simplicity, and one-size-fits-all solutions, strategies that
have lent themselves to increased legibility and thus control (270). These,
of course, contrast with forms of practical knowledge wrought not through
controlled experimentation, but through experience based on opportunity,

contingency, and expediency. While Scott talks about legibility under the apparatus of the state here, I argue that in many ways this appears overshadowed by, if not subsumed under, legibility within economic markets, even if the relationships between markets and states are extremely complex.

What is lost when a focus on the Market, and being legible within the Market, becomes all-consuming? An ethnographic example may help illuminate this. In September 2019, I attended a regional conference in Raleigh, North Carolina, that brought together southeastern Latinx farmers, many of whom were immigrants, with local extension and other agricultural professionals. Ultimately, the conference purported to celebrate the future of U.S. food and agriculture as increasingly Hispanic. Yet, the conference seemed as much geared toward the large number of extension and agricultural professionals in attendance as it did in assisting the relatively few immigrant farmers able to make the trip to Raleigh. As such, its implicit focus appeared to be educating the extension specialists and agricultural professionals in attendance on the increasing number and diverse interests of immigrant farmers and exhorting them to examine outreach deficiencies to their Hispanic farmer constituents.

In contrast, the programming geared toward the farmers took something akin to the knowledge-deficit approach Calo (2018) discusses. The afternoon panel was chaired by two Louisiana-based economic development professionals, one of whom works regularly with Latinx immigrant farmers. Their lengthy presentation constantly stressed farmers' need to own their status as businesspersons, encouraging farmers to identify themselves as "entry-level," "emerging-level," "expanding-level," or "advanced-level" entrepreneurs. Moreover, they implored farmers to identify their unique market contributions and accept that the inevitable failures would come. In these times, farmers should aim to "fail small, fail fast, pivot, and repeat."

The trainers suggested that Latinx immigrant farmers could ascend from "entry-level" toward "advanced" by becoming inscribed within, and increasingly legible to, the Market in various ways. In other words, they encouraged the farmers to document and produce increasing metrics that would define the health and standard operating procedures of their business. Such practices included generating and analyzing monthly profit and loss statements, tracking costs, producing written job descriptions, composing human resources and operations manuals, developing elevator pitches about one's business and its unique contribution, and creating an organizational chart and business succession plan, among other things. Many of these assessment tools demand literal inscription—the creation of documents, spreadsheets, and charts—that turn day-to-day farm operations into data points that make the business legible to bank personnel, FSA loan officers, or USDA staff who assess grant and program eligibility.

While this advice is certainly valuable for beginning businesses, as Calo (2018, 368) observes among the California farmer assistance programs he studied, it highlights the conference organizers' assumption that the biggest deficit of the Latinx immigrant farmers in attendance was their lack of business acumen. In other words, it was assumed that they may know how to grow, but they may not know how to run a business properly. And with this assumption comes an emphasis on farming as an economic enterprise, which envisions farmers as nascent businesspeople constantly on the hunt for greater market opportunities and increased efficiency. As Minkoff-Zern (2019) found among the farmers she studied, this is often not the case. In reality, it ignores the often overwhelming structural issues immigrant farmers face, like access to land and capital, that make or break their farming operations (Calo 2018). Furthermore, the record keeping for which these presenters advocated often goes against cultural norms in these farmers' communities of origin, where knowledge is disseminated orally and not easily recorded in writing (Minkoff-Zern 2019). Or, as Sebastian, a Latinx immigrant farmer I interviewed, noted, implementing the record keeping required was simply not something he had the time to do amid the countless other tasks operating his farm required. Unfortunately, this prevented him from obtaining the much-desired organic certification he hoped to gain.

Such a sharp focus on agriculture as business in some ways implicitly diminishes the goals, motivations, and meaning that farmers may attribute to their farming enterprises. In this same talk, the presenters at different points seemed both to extol and to diminish the value of the "lifestyle" factors to which many farmers gravitated, such as a flexible schedule and the ability to work in nature. Yet, time and again, farmers at the conference identified personal connection and meaning as what drove them to farm despite the difficulties. For example, dragon fruit grower Elisa affirmed that, although farming is hard, she feels she is "the happiest woman in the world" since, in her work on the farm, she is "with nature and not caged anymore," as she had felt in her prior job. Critically, she also noted that the freedom the dragon fruit farm gave her also connected her to memories of her great-grandparents', grandparents', and parents' coffee farm in Venezuela and their annual prayers before planting, asking for God's blessings on their and their neighbors' harvests. As Sebastian's and Elisa's words show, business acumen and detailed records may be valuable when running a farm, but they are often the last thing on a farmer's mind.

Moving from Legibility to Embodiment

The motivations, memories, and forms of meaning making, like those Elisa spoke of, do not easily lend themselves to inscription. Yet just as inscription of farm activities has become a de facto lingua franca for program and loan eligibility, so too has inscription come to dominate the world of biodiversity conservation, despite the embodied activities that have long actively preserved plant and animal varieties in situ. It is important to acknowledge that the kind of inscription and legibility discussed above—which the market demands, and which converges to make farmers businesspeople—is fundamentally different from the kind of black boxes and inscription that conventional biodiversity conservation has employed (Latour 1987; Latour and Woolgar 1986; Power 1996). Their difference lies primarily in divergent goals. Biodiversity conservation of seed germplasm relies on collecting, documenting, and executing management plans for ex situ conservation. It does so, as Nazarea (2021) points out, on the threat of loss and the need to salvage. In contrast, the market focuses on legibility for simplification and systematization that allows comparison and evaluation of those businesses deemed more fit to survive, not to mention bolstering a voluminous and profitable industry of data collection and auditing. Conservation focuses on the attempted collection and preservation of all species, while the market focuses on culling all businesses but those that are supposedly the fittest.

Nonetheless, while divergent in their aims, the inscriptive modalities of conventional conservation and the market are similar. Both rely on collecting, documenting, and organizing data into comparable, one-size-fits-all forms. Like the market-oriented application, inscription for the sake of biodiversity conservation is a counterpart to the "catechism of high-modernist agriculture" that Scott describes (1998, 270), even if the aims are different. Rather than gathering simplified data for comparison and evaluation, biodiversity conservation and documentation derives from a fear of loss (Nazarea 2006). As Huyssen (2000) argues, this trend—something he identifies as a postmodern reaction to modernist ideals—finds echo in the late twentieth-century surge of museums, nostalgia-oriented enterprises, and even individuals' desires to document their own lives through various forms of media, a tendency that has escalated even more in the era of social media. Huyssen is critical, however, of the idea that a documentary profusion can truly counter this loss. Rather, as Nazarea states, it is better to focus on "sensuous recollection in marginal niches and sovereign spaces that people carve out of uniformity and predictability" as these are better able to "[replenish] what modernity drains" (2006, 320). Small farms and gardens may not be the most publicly visible sites for the embodied preservation of biodiversity; however, the actions taking place within them are fundamentally productive for the preservation of interspecies relationships, not only for individuals but for the collective whole.

To speak of farming as sensuous practice recognizes that humans engage with environments on a fundamental embodied level. These engagements shape individual and collective identities while transmitting social memory and ecological knowledge across generations through performative acts like older generations teaching children about gardening at a young age or using gardens as spaces of recreation and creative expression (Crumley 2002). This may be especially true for immigrants. A substantial body of literature on immigrant home and community gardens paints an engaging portrait of how gardening offers immigrants creative opportunities to maintain transnational linkages while building a sense of place in new communities. As such, several scholars focus on the affective, embodied, and performative nature of gardens, which they identify as spaces of tradition and hybridity (Gerodetti and Foster 2016; Nazarea 2013).

Scholars also identify homegardens as important religious, cultural, and family memorial spaces, where gardening connects them to flora from home as well as the people and cultural traditions these environments signify (Mazumdar and Mazumdar 2012; see also Nazarea 2005). Critical here is the creativity involved in immigrants' evolving senses of place and means of maintaining transnational ties (Morgan, Rocha, and Poynting 2005). As Graham and Connell state, "gardens provide a setting for creativity, a connection to personal history, a reflection of one's identity and foster a sense of place within the broader urban environment" (2006, 379). In these creative multispecies engagements, immigrants exercise agency in how they articulate their evolving identity and hybridity (Levitt and Jaworsky 2007). Minkoff-

Zern (2019) contends that this is a critical motivation for the Latinx immigrant farmers she interviews in starting and maintaining their farms, and for involving family members, particularly their children, in farm activities.

Scholarly work on community gardens likewise points to how gardens help immigrants produce their own food, link past to the present, and build community. Saldivar-Tanaka and Krasny argue that Latinx community gardens in New York are important "participatory landscapes" that play a role in land access, advocacy, horticultural knowledge, and connections to cultural heritage (2004, 409). Hondagneu-Sotelo argues that, in addition to being important spaces for community building, Los Angeles community gardens allow Latinx immigrants to engage in "social reproductive and restorative activities." As such, she argues, they are an extension of the domestic sphere (2017, 13). Gardens are also spaces of autonomy and agency. For farmworkers in particular, gardens offer a chance to exercise traditional agricultural knowledge and provide a critical safety net in the face of food insecurity (Minkoff-Zern 2014). Moreover, they remain important reservoirs of biodiversity, polyculture, and traditional agricultural knowledge, providing opportunities for future generations to connect to these plants and gardening practices (Airriess and Clawson 1994; Andrews-Swann 2013; Imbruce 2007; Nazarea 2013).

Consequently, engaging with landscapes, plants, and the foods they produce draws on all the senses, calling forth an almost artistic means of coming to dwell in new landscapes and homes. Sensory and embodied engagements help root immigrants in new locales, even as they preserve important ties to people, landscapes, and memories from home. This is not to overromanticize or overextend the work of the sensuous, but to acknowledge its fundamentally generative nature. Intimately tied to this, I argue, are the relationships these embodied practices engender, even those stemming from market interactions. The New York farmer I spoke with, named Neto, reveals ways that embodied practice and market relationships are intimately connected. He speaks of his plants as his babies, and in the same breath, he recounts how meaningful it is that his farmers' market customers return, express appreciation for his produce, and seek his advice on growing their own gardens.

Neto: A Colombian Small Farmer in the Big City

Neto's story exemplifies well the connection between embodied practice and market inscription for immigrant farmers. He farms west of New York City and sells his products within markets in Brooklyn and Queens. As we spoke, sitting on the steps at the entrance to a Brooklyn school, I watched customers flood his popular stand, many turning to wave at him (figure 6.1). For many customers, he briefly interrupted our interview to run over and chat with them, greeting them by name and with a firm handshake or hug. As evidenced by both his words and actions, the farmers' markets at which he sells his produce have allowed him to build close relationships with his customers.

The popularity of his farmers' market stand, however, is not merely due to his affable nature. He has succeeded largely because of the more than fifty-five varieties of vegetables he grows. These include typically popular vegetables in the United States like kale and carrots, as well as herbs from his native Colombia, and naturally growing local wild edibles like purslane and amaranth, favorites among his Latin American and Caribbean customers. The many species and varieties have allowed him not only to support his business but also to engage with plant knowledges he developed as a child. This now transnational knowledge is embodied and actualized in the work Neto puts into his farm to grow his plant "babies," the vegetables that

FIGURE 6.1 Neto's farm stand in 2018 at a Saturday Brooklyn greenmarket, boasting some of the fifty-five types of vegetables he grows. Photo by author.

become the living embodiment of this knowledge. These special varieties he grows he then introduces to customers of all backgrounds and ethnicities across New York City greenmarkets. In turn, knowledge about the existence of these plants, their varieties, their culinary flavors, and how to grow them is transmitted to and becomes embodied within and among his customers.

Maria: A Pig Farmer and Peruvian Pepper Purveyor

Maria's farming practices are another example of the complex connections between sensory memory and commercial success. She was raised in an urban part of Peru, and although she's had indirect connections to farming through family friends, it was never a particular interest of hers. She moved to the United States as an adult and worked in business in California, and later in Louisiana, for many years. After retiring, she became engaged to a local Louisiana man, whose primary job was on a tugboat, but who had also begun raising heirloom pigs on his land north of Lafayette. She stumbled into farming through him and began learning to raise pigs and grow vegetables because her help was needed in his absence during two-week stints on Gulf tugboats.

Having an entrepreneurial spirit, Maria also started raising laying hens to begin an egg business. This compelled her to delve into other agricultural domains, from local Louisiana tomato varieties to a special kind of lemon available only in the north of Peru to appropriately flavor her ceviche. She remains proudest, however, of the peppers she is working to cultivate. Maria had craved fresh aji amarillo (Peruvian yellow) peppers of her native Peru; however, she could only find them frozen or canned fresh locally. As a result, she had relatives send her seeds so she could experiment with growing the plant in the hot Louisiana summers (figure 6.2). After three seasons, she had started to achieve some success and began making plans to sell these peppers in her area. Moreover, she started including them in her fiancé's renowned boudin sausage recipe, a local Cajun specialty and a value-added product they eventually plan to sell from the heirloom pigs they raise. As the aji amarillo pepper, lesser-known in the United States, makes its debut in this regional sausage, it not only adds a transnational flare to a local Cajun dish, but also helps Maria maintain and perpetuate connections with the sensory memory of flavor from her native Peru.

Again, we see how the desire to preserve and reconnect with flavor memories from home compels seeds' international travel and preservation (see

FIGURE 6.2 *Left*, some of the pepper seedlings in a tray Maria was experimenting with; *right*, Maria proudly shows off her ripe aji amarillo peppers, which she nurtured from seed in the hot Louisiana climate. Photos by author.

Anastario, this volume). In Maria's case, however, while the flavor memories persisted from her youth, her plant-based knowledge was not resurrected but rather developed, cultivated, and embodied in her new home. Although the memories motivating her to seek out the aji amarillo pepper are initially confined to one domain—here, Maria's kitchen—the know-how to actualize this pepper's in situ preservation required Maria to expand into the unfamiliar domain of agriculture to build new embodied knowledge on plants and their cultivation. The aji amarillo pepper plants, in turn, are adapting to a new latitude and climate, while making their debut among new consumers, who then develop their own flavor memories.

This example points to the multiple levels at which immigrant farmers like Maria are helping to nurture the embodiment of biodiversity, not only for themselves but for those with whom they share their agricultural craft. As for Neto, these shared flavors and knowledges of plants, perhaps previously unfamiliar to Maria's clientele, are wrought through markets. Nonetheless, they derive not from intense rational planning and documentation of farm activities, but rather from the pursuit of creative experimentation driven by memory and taste.

José and Family: Actualizing Their Vision for Medicinal Herbs

Like Neto and Maria, José is another farmer espousing a clear vision for his business that relies on deeply embodied practices. On a former strawberry farm near Palm Beach, Florida, that he and his family recently took over, they run a produce stand and cultivate several acres. In 2018 this land appeared to be an almost indistinguishable mass of weeds and abandoned pots that had been left for the ever-creeping South Florida vegetation to take over. Yet out of this seeming disorder, their self-defined vision of a "farm-acy" was coming to life. Although José is originally from Uruguay, he spent several years in Peru learning to become a shaman. He uses this shamanic knowledge to create herbal supplements that he sells to the local community from the fresh and dried organic medicinal herbs and plants he and his family grow, especially moringa trees and seeds. *Moringa oleifera* is a tree native to the Indian subcontinent that grows well in tropical and subtropical areas, whose leaves, bark, flowers, fruit, seeds, and root are traditionally used for ailments ranging from anemia and thyroid disorders to stomach disorders and low sex drive (Razis, Ibrahim, and Kntayya 2014). During my visit, a steady stream of customers—mostly Latinx women—waited their turn for a consultation with José, where he would assess their conditions and then prescribe appropriate herbal remedies, many of which use moringa. José's skill is remarkable; I witnessed him make spot-on inferences about the healthfulness of a person's diet simply by looking at their face and skin.

José's skill as an herbalist and shaman in many ways is paradigmatic of the deeply embodied knowledge he possesses and practices. He knows intimately the plants and the properties each part of the plant contains—leaves, roots, seeds, stems—and utilizes this knowledge to create herbal supplements and tinctures that treat various ailments, many of which he links to how contemporary life confronts us with unhealthy activities and foods and has divorced us from natural remedies. For those that seek his advice and remedies, his knowledge and the plants he draws from in turn act on and become part of the bodies of those that take them, hopefully correcting their insalubrious state. Thus, while the cultivation that José, his wife, and daughters engage in are embodied practices in and of themselves—given how they engage physically with the soil, plants, and their products—his knowledge on the use of these plants to make remedies becomes wrought into the bodies of those seeking them.

The family's practices go well beyond a simple herbal remedy focus. He and his family also run a farm stand, where they sell their own organic plant starts and fresh eggs, as well as reselling locally grown vegetables (figure 6.3). Over time, José and his family have also expanded to selling body products like loofahs and organic herbal body scrubs they have developed, all of which are available on their Facebook marketplace site. One of José's adult daughters, Delia, is foundational in running the farm stand and advising customers on the care and use of the plants they buy. She is equally passionate about the family's vision for a farm-acy; Delia is a cancer survivor who credits her recovery, in part, to healthy eating and herbal remedies. A mother to two young children, she is active in using the farm to teach her own and other local children about gardening and where food comes from. She has even been working on an illustrated children's book about her oldest daughter's adventures on the farm to share these experiences with more children. As a result, a wide variety of plant and farm knowledge is being passed not only to adult customers but even to the very young. Collectively, José, Delia, and the rest of their family have a well-developed vision for using plants to promote human health. Their many activities draw together a diverse set of local people who benefit from their profound—and profoundly embodied—herbal knowledge and their vision of providing healthy, organic foods to South Florida.

Sebastian: Using Permaculture to Improve Local Food Security

For Sebastian, the Ecuadorean farmer who described farming as "a system of getting into debt," agriculture's sensuous embodiment is written into the landscape. Reading landscapes and plants' roles in them is a skill he learned after taking a permaculture course in Homestead, Florida. This course not only acquainted him with the agricultural principles he now espouses in all his farm activities, but it also connected him with his farm business partner, Jonathan, and a host of other like-minded agriculturalists in the

FIGURE 6.3 *Top*, potted plants and moringa trees amid the creeping Florida vegetation on the farm of José and his family; *bottom*, their farm stand—note the bottles of José's many herbal remedies in the glass case. Photos by author.

Miami-Dade region. Interestingly, small-scale organic fruit and vegetable production makes up a relatively small but growing agricultural sector across Miami-Dade and Broward Counties, in part because of the dominance of large-scale fall-winter vegetable agriculture, nursery businesses, and tropical fruit and avocado orchards in the area. Of note, I observed conventional farmers not infrequently appearing at odds with one another; however, local organic growers—comprising both white and Latinx farmers (a few of them immigrants like Sebastian)—generally know and work with one another collegially to spread their vision for healthy and sustainable local food.

Sebastian asserts that his permaculture knowledge has allowed him to link his identity more concretely to his family's farming roots in Ecuador, a realization that he feels has helped him in turn become more comfortable with his life in the United States. It has also helped him facilitate communication with his grandmother, a conventional farmer in Ecuador, about ways to make their farm there more environmentally sustainable. He expressed his sentiments about the importance of agriculture:

> To already have been part of that community, coming here, and being in a completely different alien land to me, where I had to relearn a new different lifestyle of speaking, of living, of communicating, you know . . . once I started back into agriculture, it kind of drew me through. After doing that, it just kind of started to sink in: "Man, this is what my family does, what my family did, what my family grew up in." So now, I can communicate with my grandma about farm stuff.

His permaculture focus also intimately connects with the values he has anchored his business in. Sebastian had previously grown fruits and vegetables for sale to local markets and Community Supported Agriculture (CSA) programs. When the land he and several others rented to farm was sold, however, he and Jonathan decided to shift gears. On the four acres they now rent, they have opened a nursery for edible perennials, which they market largely to community homeowners aiming to grow their own gardens. Sebastian identifies a larger purpose, though. As he puts it:

> Our main mission is to get people to grow food. For themselves. It's almost counterintuitive as a business model, because eventually you're running yourself out of the game, because everyone is doing their own

thing. It's not really like that, though. In the end the goal is to get more people to do this on their own. Food security is a big issue. In any county, but especially in Broward County. We want to make sure Broward County is food secure. . . . I don't want to be the person who, if I'm not there, nobody eats. We're trying to teach you what it takes to grow food in South Florida.

What is interesting is how Sebastian orients his personal values alongside his recognition of what it takes to be legible within markets. On the one hand, he calls his business model "counterintuitive." On the other hand, he notes that his nursery-oriented business plan makes financial sense because of the need for fewer inputs and less infrastructure, which in effect makes him more appealing to loan officers and potential landlords. This in turn allows him to grow what is meaningful to him: a "permaculture fruit forest" with coconuts, bananas, lychee, coffee, and a wealth of other edibles like Tahitian spinach, moringa, purslane, cranberry hibiscus, yucca, holy basil, and pineapples. "When these are fully grown," he states, "you've got a beautiful, diverse, insect- and pollinator-attracting guild of plants that produce a ton of food." Here, we see Sebastian actively framing his sphere of embodied practice, sedimented from memories of plants and tastes from his native Ecuador, alongside the permaculture practices he learned stateside. He does this in ways that help him gain legitimacy within external institutions.

By the time I returned in 2020, Sebastian's farm business had really begun to take shape. Where there had been a single high tunnel in 2018 were now two side by side: one devoted to the plants, organic soil amendments, and garden supplies for sale, and the second for starting the seedlings that they sold to customers to create edible gardens. The selection of plants, too, had blossomed. Sebastian and Jonathan were adamant about not offering plants to customers that would fail out of season, like tomatoes that would never survive a humid South Florida summer. Instead, they offered a breadth of ready-to-plant vegetables like collards, kale, chard, radishes, onions, and several varieties of peppers and sorrel; culinary herbs like stevia, rosemary, nasturtiums, fennel, cilantro, chives, and several kinds of basil and mint; and medicinal herbs like *Spilanthes*, motherwort, marigold, and bee balm. Across from their high tunnels was a sizable garden plot run by some fellow organic farmers who sold CSA shares locally. Additionally, to the front, their landlord was helping to construct a pavilion in which Sebastian and

Jonathan would host regular weekend farmers' markets, bringing together several other local organic farmers and a Latina soil scientist—herself an immigrant—who develops and sells her own compost and soil mixes.

Not surprisingly, COVID-19 brought many of these plans to a screeching halt. The first farmers' market planned for mid-March 2020 was preemptively canceled out of "an abundance of caution" because of increasing coronavirus spread. When I next saw him, Sebastian lamented the loss of the expected five thousand dollars in revenue the market was slated to bring. Just days later, however, as news of the coronavirus's spread was initiating the shutdown of schools and workplaces, Sebastian and Jonathan's fortunes had turned. They pivoted their business overnight, loading their entire stock of plants to their Weebly website and point-of-sale software. They then created an online store and sent out email blasts to all their customers to advertise their new curbside pickup system. In turn, one customer shared their website on Facebook, resulting in a flurry of new customers, including one who bought nine hundred dollars' worth of plants, soil, and garden supplies. Plants that Sebastian had anticipated having to dispose of since he had planted too many were sold out within hours. That afternoon and into the evening, we packed box after box of prepaid orders (figure 6.4). From a safe social distance, we entertained the few customers who requested help in selecting plants to start their gardens amid the increasing supply chain issues created early in the pandemic. Reflecting on the day, Sebastian and Jonathan noted how satisfied they felt being able to contribute to local food security when supply chains seemed so uncertain. The also reflected on how good it was to see young families actively getting children involved in gardening, spreading knowledge about how food grows.

Perhaps more than most farmers I talked to, Sebastian seemed oriented to the details that would make his and Jonathan's farm business financially sustainable. He described his pointed focus on creating an effective business plan from the outset and considered himself more the detail-oriented "production guy," while Jonathan's big-picture focus made him better suited to marketing their business and building a social media following. Social media became a lifeline connecting them to customers during the early pandemic, allowing them to keep their farm afloat at a time when many businesses suffered greatly and were beginning to shutter their doors. This was the case not only for their farm, but for many other small organic farms across South Florida, including José's. In this way, social media—even if it is another form

FIGURE 6.4 *Left*, plant stands filled with young plants in Sebastian and Jonathan's high tunnel, where customers (prepandemic) seek advice about what to plant; *right*, boxes filled with plants made from an online order at the start of the COVID-19 pandemic. Photos by author.

of legibility, albeit one that often rests less on farm metrics and more on narratives or photos that draw on meaning and memories—seems to have a somewhat democratizing effect in its ability to reach customers in small *m* markets, those reflective of a diverse and complex set of relationships and exchanges. This is especially true for those able to harness its power effectively to build and hone relationships with customers.

As much as Sebastian maintained an active awareness of the business side of things, a skill he noted his father had helped him cultivate, he was equally, if not more, aware of the relational side of his business. He frequently recounted how he considered cultivating relationships to be one of the cornerstones of their farm. He described the relationship he had built with a Jamaican farmer who built him a plant stand. Sebastian, in turn, reciprocated to the Jamaican farmer with a cost break on organic fertilizer or new plant varieties when possible. Likewise, he found it important both to be honest with customers when he did not have what they requested and to make referrals to other nurseries—in turn building relationships with them—so that the customers found what they sought. Making sure he assisted customers immediately on arrival was critical, since many customers needed lengthy consultations to decide on their purchases. During one particularly busy day, Sebastian noticed—for the first time in three years—one customer who got impatient and left. He relayed, "It just killed me. We rely on word of mouth from our customers to grow." To facilitate the excellent customer service he strove to provide, Sebastian was working on a detailed labeling system for his plant starts. This would allow customers to read about the varieties they were looking at and begin to make decisions while he helped others. He also seemed acutely aware of the need to help customers, but not to the extent that they "used" him for his knowledge and never returned. Rather, his goal was to build relationships so that customers would "invest in [his] service" and return in the future.

Farming as a Balancing Act

Stories like those recounted above help us reconsider how market dominance and legibility within the Market is framed. So often, the Market is portrayed as a monolith, an entity that deterministically wields power. This is something captured succinctly in J. K. Gibson-Graham's term "capitalocentrism," describing contexts in which capitalism is so foregrounded that it results in the diminution or marginalization of activities deemed less or noncapitalist (1996, 41). Likewise, in the example of the conference above, actions that support a seemingly singular Market rationality, establish efficiency, and make farmers legible to banks or government agencies are heralded as best, and other activities or pursuits are marginalized. Meaning, memory, and the often embodied pursuits that actualize these frequently seem to fall short, unless they readily lend themselves to activities deemed orthodox by the Market.

Instead, it is better to speak of inscription across multiple markets and at multiple scales, bringing together farmers and consumers in various ways. As Subramaniam and Willey (2017) speak of the difference between "Science" as the legitimizing apparatus of institutions and "sciences" as the multiple knowledges that exist at the margins, the small *m* markets in which I saw many of these immigrant farmers engage vary in their degree of legibility and legitimization. This again coincides well with the perspective of Gibson-

Graham and allied authors, who focus on a "diverse economy" supported by activities normally associated with capitalism, as well as a wide variety of nonmarket and alternative market transactions, forms of labor, modes of compensation, enterprise forms, and modes of accumulation and distribution (St. Martin, Roelvink, and Gibson-Graham 2015, 4). To achieve this perspective requires an ontological shift to avoid "seemingly fixed and stable essences" and to reframe these into "successful processes of becoming" (6). Focusing on processes of becoming and multiple and diverse incarnations of economic activity allies particularly well with the kind of creative multifocal projects the Latinx farmers I met undertook, as well as the processual and embodied nature of these many activities.

These perspectives bring these farmers together with other actors in ways that help to support the personal or value-based choices they make on their farms and perpetuate the sensuous, embodied practices in which they engage. While it is possible to view these market interactions as hierarchically scalar, from localized to more impersonal exchanges, viewing them instead as different sites or analytical fields (Holloway et al. 2007) can point us more to the relationships created in these exchanges, and the embodied experiences they indicate. Inscription in markets is multivalent; the inscription of farm activities and mobilization of customer interest through stories and photos on social media provides one such example, illustrated in several of the vignettes above, of how inscription facilitates market interactions. Nonetheless, this relies more on relational interactions rather than legibility through metrics and statistics. While this necessitates a different skillset of creative marketing and branding, it draws often on the very memories and meaning undergirding embodied practice. Thus, relationships and values, and the extent to which Latinx immigrant farmers mobilize them to gain or to maintain legitimacy in various markets, can help to create spaces in which Latinx immigrant farmers are able to pursue meaningful projects that, in turn, embody and actualize biodiversity.

Following Calo (2018), Minkoff-Zern (2019), Minkoff-Zern and Sloat (2017), and others, focusing on the structural limitations Latinx immigrant farmers face as they start and maintain their farming operations is essential. Doing so is necessary to challenge dominant narratives and bring about change in local and global agrifood systems, given the immense role markets play in levying many of these structural inequities. At the same time, though, and as I do in this chapter, it is equally valuable to highlight and celebrate the creativity that these agriculturalists employ to make their farming operations successful. This includes focusing on the practices in which they engage both on the land (Minkoff-Zern 2017, 2019) and in the marketplace (Schmid 2020). The farmers above engage in a creative balancing act, making themselves visible in markets while engaging in meaningful agricultural practices and operations that allow them to contribute in small ways to biodiversity conservation. So too must scholars work to walk this fine line between critical perspectives that challenge the status quo and blithely optimistic ones that risk essentializing those we study and potentially (inadvertently) perpetuating problematic power structures.

It is also important to acknowledge the valid concern raised by Korsunsky (2020) about the unchecked romanticization inherent in portraying Latinx immigrant farmers as bastions of sustainable agricultural practices and paragons of food sovereignty. Although the farmers I describe above do fit Minkoff-Zern, Welsh, and Ludden's (2019) definition of those engaged in alternative farming practices—an indeterminate set of practices typically with a sustainable and community and food security focus that lie outside standard agro-industrial practices—the examples I raise here are not meant to suggest only that. Several farmers I met farmed conventionally, and not all those I met explicitly sought to preserve biodiversity as a primary, or even secondary, aim. U.S. agriculture is as diverse a landscape in its praxis as is the biodiverse germplasm that composes and sustains it. Thus, it may be valuable to consider how the practices of Latinx immigrant farmers might parallel the current variegated landscape of U.S. agriculture. Just as gardeners, small farmers, and increasingly organizations engage in small but progressively visible efforts to promote biodiversity, sustainability, and food sovereignty, so too do many Latinx farmers. These endeavors occur through various activities, many of which include holding on to the plants that tie them to memories of other people, places, and times. Considered against the market orientation and presumed economic rationality that has driven U.S. agriculture for nearly a century now, these efforts surely seem to lie at the margins. When considered as a whole, however, these efforts and the kinds of embodied biodiversity they enact are exceedingly valuable.

Ultimately, I believe the examples shared in this chapter help to reconceptualize the wide variety of ways in which farmers engage with "the Market" despite the Market's seemingly narrow need to legitimize farmers primarily by their creation of business models and marketing strategies. Rather, looking at such examples of small *m* markets helps scholars and practitioners understand how Latinx immigrant farmers are simultaneously able to perpetuate embodied and sensuous engagement with meaningful landscapes and plants, build relationships with customers, and nurture businesses that often reflect deeply held moral visions. As transnational brokers of sometimes unfamiliar crops and flavors, a not insignificant number of Latinx immigrant farmers continually work to connect their livelihood to sensuously embodied craft and often do so with success. This does not come without hard work, but it seems to align especially closely with the motivations and memories that compel their experimentation and creativity.

References

Airriess, Christopher A., and David L. Clawson. 1994. "Vietnamese Market Gardens in New Orleans." *Geographical Review* 84 (1): 16–31.

Andrews-Swann, Jenna E. 2013. "Exile Landscapes of Nostalgia and Hope in the Cuban Diaspora." In *Seeds of Resistance, Seeds of Hope: Place and Agency in the Conservation of Biodiversity*, edited by Virginia D. Nazarea, Robert E. Rhoades, and Jenna E. Andrews-Swann, 240–61. Tucson: University of Arizona Press.

Boyd, William, and Michael Watts. 1997. "Agro-Industrial Just-in-Time: The Chicken Industry and Postwar American Capitalism." In *Globalising Food: Agrarian Questions and Global Restructuring*, edited by David Goodman and Michael Watts, 192–225. London: Routledge.

Burns, Christopher, and Ryan Kuhns. 2016. *The Changing Organization and Well-Being of Midsize U.S. Farms, 1992–2014*. Economic Research Report No. 219 (ERR-219). U.S. Department of Agriculture, Economic Research Service, October. https://www.ers.usda.gov/webdocs/publications/80692/err-219.pdf?v=0.

Calo, Adam. 2018. "How Knowledge Deficit Interventions Fail to Resolve Beginning Farmer Challenges." *Agriculture and Human Values* 35 (2): 367–81.

Calo, Adam, and Kathryn Tiegen De Master. 2016. "After the Incubator: Factors Impeding Land Access Along the Path from Farmworker to Proprietor." *Journal of Agriculture, Food Systems, and Community Development* 6 (2): 111–27.

Cochrane, Willard W. 1993. *The Development of American Agriculture: A Historical Analysis*. Minneapolis: University of Minnesota Press.

Crumley, Carole L. 2002. "Exploring Venues of Social Memory." In *Social Memory and History: Anthropological Perspectives*, edited by Jacob Climo and Maria G. Cattell, 39–52. Walnut Creek, Calif.: AltaMira.

DeGuzmán, María. 2019. "LatinX Botanical Epistemologies." *Cultural Dynamics* 31 (1–2): 108–24.

Fitzgerald, Deborah K. 2003. *Every Farm a Factory: The Industrial Ideal in American Agriculture*. New Haven, Conn.: Yale University Press.

Flora, Jan L., Mary Emery, Diego Thompson, Claudia M. Prado-Meza, and Cornelia B. Flora. 2011. "New Immigrants in Local Food Systems: Two Iowa Cases." *International Journal of Sociology of Agriculture and Food* 19 (1): 119–34.

Garcia, Victor, Bernardo Lopez-Ariza, and Juan Marinez. 2008. "Exploring Undercounts in the Agriculture Census: An Alternative Enumeration of Hispanic Farmers in Southwestern Michigan." *Journal of Latino/Latin American Studies* 3 (2): 47–58.

Garcia, Victor, and Juan Marinez. 2005. "Exploring Agricultural Census Undercounts Among Immigrant Hispanic/Latino Farmers with an Alternative Enumeration Project." *Journal of Extension* 43 (5): 1–6.

Gerodetti, Natalia, and Sally Foster. 2016. "'Growing Foods from Home': Food Production, Migrants, and the Changing Cultural Landscapes of Gardens and Allotments." *Landscape Research* 41 (7): 808–19.

Gibson-Graham, J. K. 1996. *The End of Capitalism (as We Knew It): A Feminist Critique of Political Economy*. Minneapolis: University of Minnesota Press.

Graham, Sonia, and John Connell. 2006. "Nurturing Relationships: The Gardens of Greek and Vietnamese Migrants in Marrickville, Sydney." *Australian Geographer* 37 (3): 375–93.

Hightower, Lisa S., Kim L. Niewolny, and Mark A. Brennan. 2013. "Immigrant Farmer Programs and Social Capital: Evaluating Community and Economic Outcomes through Social Capital Theory." *Community Development* 44 (5): 582–96.

Holloway, Lewis, Moya Kneafsey, Laura Venn, Rosie Cox, Elizabeth Dowler, and Helena Tuomainen. 2007. "Possible Food Economies: A Methodological Framework for Exploring Food Production–Consumption Relationships." *Sociologia Ruralis* 47 (1): 1–19.

Hondagneu-Sotelo, Pierrette. 2017. "At Home in Inner-City Immigrant Community Gardens." *Journal of Housing and the Built Environment* 32 (1): 13–28.

Huyssen, Andreas. 2000. "Present Pasts: Media, Politics, Amnesia." *Public Culture* 12 (1): 21–38. https://doi.org/10.125/08992363-12-1-21.

Imbruce, Valerie. 2007. "Bringing Southeast Asia to the Southeast United States: New Forms of Alternative Agriculture in Homestead, Florida." *Agriculture and Human Values* 24:41–59.

Korsunsky, Alex. 2020. "Back to the Root? Immigrant Farmers, Ethnographic Romanticism, and Untangling Food Sovereignty in Western Oregon." *Culture, Agriculture, Food, and Environment* 42 (2): 114–24.

Latour, Bruno. 1987. "Introduction: Opening Pandora's Black Box." In *Science in Action: How to Follow Scientists and Engineers through Society*, 1–17. Cambridge, Mass.: Harvard University Press.

Latour, Bruno, and Steve Woolgar. 1986. *Laboratory Life: The Construction of Scientific Facts*. Princeton, N.J.: Princeton University Press.

Levitt, Peggy, and Bernadette Nadya Jaworsky. 2007. "Transnational Migration Studies: Past Developments and Future Trends." *Annual Review of Sociology* 33 (1): 129–56. https://doi.org/10.1146/annurev.soc.33.040406.131816.

Lewontin, Robert C. 2000. "The Maturing of Capitalist Agriculture: Farmer as Proletarian." In *Hungry for Profit: The Agribusiness Threat to Farmers, Food, and the Environment*, edited by Fred Magdoff, John Bellamy Foster, and Frederick H. Buttel, 93–106. New York: Monthly Review.

Mazumdar, Shampa, and Sanjoy Mazumdar. 2012. "Immigrant Home Gardens: Places of Religion, Culture, Ecology, and Family." *Landscape and Urban Planning* 105 (3): 258–65.

Minkoff-Zern, Laura Anne. 2014. "Hunger Amidst Plenty: Farmworker Food Insecurity and Coping Strategies in California." *Local Environment: The International Journal of Justice and Sustainability* 19 (2): 204–19.

Minkoff-Zern, Laura Anne. 2017. "Crossing Borders, Overcoming Boundaries: Latino Immigrant Farmers and a New Sense of Home in the United States." In *Food Across Borders*, edited by Matt Garcia, E. Melanie DuPuis, and Don Mitchell, 219–35. New Brunswick, N.J.: Rutgers University Press.

Minkoff-Zern, Laura Anne. 2018. "Race, Immigration and the Agrarian Question: Farmworkers Becoming Farmers in the United States." *Journal of Peasant Studies* 45 (2): 389–408.

Minkoff-Zern, Laura Anne. 2019. *The New American Farmer: Immigration, Race, and the Struggle for Sustainability*. Cambridge, Mass.: MIT Press.

Minkoff-Zern, Laura Anne, and Sea Sloat. 2017. "A New Era of Civil Rights? Latino Immigrant Farmers and Exclusion at the United States Department of Agriculture." *Agriculture and Human Values* 34:631–43.

Minkoff-Zern, Laura Anne, Rick Welsh, and Maizy T. Ludden. 2019. "Immigrant Farmers, Sustainable Practices: Growing Ecological and Racial Diversity in Alternative Spaces." *Agroecology and Sustainable Food Systems* 44 (7): 947–72.

Morgan, George, Cristina Rocha, and Scott Poynting. 2005. "Grafting Cultures: Longing and Belonging in Immigrants' Gardens and Backyards in Fairfield." *Journal of Intercultural Studies* 1 (2): 93–105.

Nazarea, Virginia D. 1998. *Cultural Memory and Biodiversity*. Tucson: University of Arizona Press.

Nazarea, Virginia D. 2005. *Heirloom Seeds and Their Keepers: Marginality and Memory in the Conservation of Biological Diversity*. Tucson: University of Arizona Press.

Nazarea, Virginia D. 2006. "Local Knowledge and Memory in Biodiversity Conservation." *Annual Review of Anthropology* 35:317–35.

Nazarea, Virginia D. 2013. "Temptation to Hope: From the 'Idea' to the Milieu of Biodiversity." In *Seeds of Resistance, Seeds of Hope: Place and Agency in the Conservation of Biodiversity*, edited by Virginia D. Nazarea, Robert E. Rhoades, and Jenna E. Andrews-Swann, 19–41. Tucson: University of Arizona Press.

Nazarea, Virginia D. 2021. "Ontologies of Return: Terms of Endearment and Entanglements." In *Moveable Gardens: Itineraries and Sanctuaries of Memory*, edited by Virginia D. Nazarea and Terese V. Gagnon, 253–72. Tucson: University of Arizona Press.

Niewolny, Kim L., and Patrick T. Lillard. 2010. "Expanding the Boundaries of Beginning Farmer Training and Program Development: A Review of Contemporary
Initiatives to Cultivate a New Generation of American Farmers." *Journal of Agriculture, Food Systems, and Community Development* 1 (1): 65–88.

Power, Michael. 1996. "Making Things Auditable." *Accounting, Organizations and
Society* 21 (2–3): 289–315.

Razis, Ahmad Faizal Abdull, Muhammad Din Ibrahim, and Saie Brindha Kntayya.
2014. "Health Benefits of *Moringa oleifera.*" *Asian Pacific Journal of Cancer Prevention* 15 (20): 8571–76.

Rissing, Andrea. 2016. "Alternative Economic Strategies and the Technology Treadmill: Beginning Vegetable Farmers in Iowa." *Economic Anthropology* 3 (2): 304–14.

Saldivar-Tanaka, Laura, and Marianne E. Krasny. 2004. "Culturing Community Development, Neighborhood Open Space, and Civic Agriculture: The Case of Latino
Community Gardens in New York City." *Agriculture and Human Values* 21 (4):
399–412.

Schmid, Mary Elizabeth. 2019. "Tomatero Circuit of Southern Appalachia–South
Florida." *Anthropology of Work Review* 11 (1): 25–35.

Schmid, Mary Elizabeth. 2020. "Enterprising Women of Mexican American Farming
Families in Southern Appalachia." In *The Immigrant-Food Nexus: Borders, Labor,
and Identity in North America*, edited by Julian Agyeman and Sydney Giacalone,
161–80. Cambridge, Mass.: MIT Press.

Scott, James C. 1998. *Seeing like a State: How Certain Schemes to Improve the Human
Condition Have Failed.* New Haven, Conn.: Yale University Press.

St. Martin, Kevin, Gerda Roelvink, and J. K. Gibson-Graham. 2015. "An Economic
Politics for Our Times." In *Making Other Worlds Possible: Performing Diverse
Economies*, edited by Gerda Roelvink, Kevin St. Martin, and J. K. Gibson-Graham,
1–25. Minneapolis: University of Minnesota Press.

Subramaniam, Banu, and Angela Willey. 2017. "Introduction: Feminism's Sciences."
In "Science Out of Feminist Theory: Part One," special issue, *Catalyst: Feminism,
Theory, Technoscience* 3 (1): 1–23.

Swisher, M. E., Mark Brennan, and Mital Shah. 2007. *Hispanic-Latino Farmers and
Ranchers Project.* University of Florida Center for Organic Agriculture, Gainesville, September 2006–September 2007. https://nifa.usda.gov/sites/default/files
/asset/document/hispanic_full_report.pdf.

U.S. Department of Agriculture. 2014. "Preliminary Report Highlights: U.S. Farms
and Farmers." *2012 Census of Agriculture.* Washington, D.C.: U.S. Department
of Agriculture. https://www.agcensus.usda.gov/Publications/2012/Preliminary
Report/Highlights.pdf.

U.S. Department of Agriculture. 2019. "Farm Producers." Highlights, *2017 Census
of Agriculture.* Washington, D.C.: U.S. Department of Agriculture. https://www
.nass.usda.gov/Publications/Highlights/2019/2017Census_Farm_Producers
.pdf.

U.S. Department of Agriculture National Agricultural Statistical Service. 2019. "Hispanic Producers." Highlights, *2017 Census of Agriculture*. Washington, D.C.: U.S. Department of Agriculture. https://www.nass.usda.gov/Publications/Highlights /2019/2017Census_Hispanic_Producers.pdf.

Vásquez-León, Marcela. 2009. "Hispanic Farmers and Farmworkers: Social Networks, Institutional Exclusion, and Climate Vulnerability in Southeastern Arizona." *American Anthropologist* 111 (3): 289–301.

Wells, Miriam J. 1990. "Mexican Farm Workers Become Strawberry Farmers." *Human Organization* 49 (2): 149–56.

Wells, Miriam J. 1996. *Strawberry Fields: Politics, Class, and Work in California Agriculture*. Ithaca, N.Y.: Cornell University Press.

CHAPTER SEVEN

Biodiversity Is Alive, Biodiversity Is Lost

Memories of an Ethiopian Plant from Homegardens to the Green Legacy Campaign

VALENTINA PEVERI

> It takes a certain kind of craziness to love all that is doomed to perish.
> —DEBORAH BIRD ROSE, "IN THE SHADOW OF ALL THIS DEATH"

By the time my book *The Edible Gardens of Ethiopia* entered the final stages of production, in spring 2020, what would soon be declared as a pandemic took over and infiltrated our daily lives and ordinary ways of being, of being with others, and of inhabiting time-space. The book was and is to me a labor of love that had already survived several other pandemic episodes before the global one—passing through times of personal and professional turmoil; through the ebbs and flows of writing and rewriting; and indeed through the paradoxical longing for the physicality of those subjects that the writing aimed to translate. Not a single day went by that I did not fantasize about the everyday poetics of togetherness I had so compellingly experienced in the field. And yet, a planned detachment from fieldwork—from commingling with plants and people, and dwelling in homegardens—allowed me to quietly sit in my armchair, and to code and decode the bounty of primary data I had collected for over ten years.

The last time I set foot in one of those Hadiyya gardens, redolent of alive and abundant biodiversity, was March 2015. I was planning to return for restitution of the book to the farmers, and for pure pleasure, soon after the publication. But I still sit here in my armchair, in what is not the writer's self-imposed confinement anymore, and I feel overwhelmed by the vividness and warmth of those memories.[1] I feel even more so at times when physical contact has stopped or has become highly exclusive; when we learn to communicate via new means—sometimes across balconies, but mostly through the looking glass of a computer screen; when the social is being increasingly reformulated through the virtual.[2] The number of webinars devoted to identifying and trusting channels and solutions we could never have imagined in ordinary times, or alternative networks and entry points to be activated around research in times of self-confinement, only helped me make painfully clear a point in relation to the subjects of my ethnographic quest: that none of that research would now be possible, by any stretch of the imagination.

This impossibility is the result of both the sensory nature of the ethnographic work I conducted and of the illegibility and secrecy of the homegardens I write about, once more and with fits of nostalgia, in what follows. Sitting in my armchair, I read again and again the chapter of J. Hartigan on how to interview a plant (2017, 253–81). How can we persist in the exercise of regarding other-than-human beings, plants among them, as ethnographic subjects? Of rendering to the best of our human means their intelligence, plasticity, and social worlds? Of providing perceptual descriptions and even "becoming with" them in times when we are prevented from indulging in full sensory awareness and active attention, and from dwelling with plants where they are rooted?

My interaction with plants and people in homegardens of southwestern Ethiopia took the form of affective and bodily encounters. I have conducted my long-term ethnography in a landscape that local people—who care for the backbone of their gardens (a perennial plant called ensete) through daily connection, investment, and entanglement (or meshwork)—consider in many ways sentient, and at times even enchanted. For them, the perennial plant has soul, flesh, and traits of humanity. I have learned from local farmers to dwell not only with that specific plant, but with the complex landscape of which the plant is the beating heart. From a social science perspective, the plant has progressively taken the contours of a living instance of plant personhood (see Peveri 2020; cf. Lounela, Berglund, and Kallinen 2019; van der Veen 2014). Over time, I also had to acknowledge I had fallen in romantic love with that vegetal person (and with all its intricate socioecological ramifications), and that such love should be taken *seriously*—in the sense meant by J. Soleil Archambault (2016) when she frames human-plant relations in Mozambican private gardens: that is, literal love and not only one that anthropomorphizes other-than-human persons; one that considers aesthetics beyond utility and yet avoids a purely decorative approach to the plant world; one that refrains from the urge to own or commodify, but rather inspires different ways of relating and becoming intimate. It was through the multisensorial that experience and meaning, beyond mere knowledge, were fixed in memory and later in writing. Now a quite rare and poetic approach to evoke, back then I learned by being and breathing in the same space (often the homegarden or the kitchen) as other people and plants; and by sharing with them surfaces, sleep, and food. It was palpable intimacy. It was visceral transfer of knowledge and performance—gut to brain through hands. As beautifully put by S. Alaimo,

> Connection, interrelation, and intersubjectivity are the ontological conditions from which new delights and new ethics emerge. Pleasure spirals through these ethical ontologies that are unmistakably material rather than abstract, disembodied principles. . . . [M]emory itself is woven into the walls, history is corporeal, and sensuality becomes a practice and a praxis. These dwellings arise from the dream of an unmediated relation between human and nature; the walls do not contain, they bestow. (2016, 26)

Once the possibility of analog encounters is erased, including those with other-than-human subjects, so too their immediacy, detail, intimate texture, and vibrancy slowly evaporate. Yet, the discomforting idea that my fieldwork would not be replicable at present, in an online environment or through flat monitor screens, is evidence of how deeply embodied the biodiversity of homegardens is—how alive and place based they are, as well as resistant to quick forms of naming, inventory, and capture.

In one way, therefore, fading memories and the concomitant (deceptive) optimism in the digital turn—when online interactions are inadvertently becoming a norm rather than pandemic anomaly—emerge as a momentous time to reflect on different versions of biodiversity that populate the Ethiopian scenario from the micro to the macro scales. The COVID-19 crisis has in many ways created the conditions to find cracks for comparing and contrasting what biodiversity looks and tastes like in places where it is embraced through all senses in the everyday; and what, on the contrary, biodiversity comes to entail when it is observed in mediated viewing and framed in policy and management plans at the national level.

Whereas local farmers nurture clusters of biodiversity without naming it as such, and their narratives are indeed replete with the materiality of gardening—of scent, touch, and taste—what informs public mandates and initiatives is erasure of the visceral from the conceptual, the uprooting of the botanical from physicality, and a peculiar rendering of in vitro biodiversity, which is then mainly boosted and disseminated through institutional blogs and enthusiastic tweets. This latter narrative (and line of action) ignores spatial contexts wherein nature-human associations create zones of inter-mingling marked by embodied experiences in the everyday, and remains impermeable to plural and resilient ways of "doing" biodiversity.

This discussion places itself in the articulate debate around ecological conservation and restoration, and how these nature-based solutions can achieve sustainable goals (Perfecto, Vandermeer, and Wright 2009). The rhetoric around these actions—only rarely captured at the household, farm, or plot level and rather mobilized at the disembodied level of the global (the view from nowhere)—has intensified in recent years and particularly now, at the eve of the UN Decade on Ecosystem Restoration (2021–30). Although laudable in the abstract, such an ambitious agenda only marginally frames its goals and strategies by accounting for the social and political dimensions of restoration; nor is it informed by historical awareness of local settings and actors, and the meanings those actors attribute to landscapes; and even less seems to actively plan for the inclusion of plural voices, values, and situated knowledges (Aisher and Damodaran 2016). The typical format through which these grand plans for restoration are disseminated to the wider audience is the visually and numerically appealing one of planting impressive numbers of (generally defined) trees along with other forms of quick and eye-catching "greening" actions.

Restoration occurs whenever the need is felt and shared among stakeholders of assisting the recovery of an ecosystem that has been degraded, damaged, or destroyed; whereas many state as well as nonstate actors "act at a distance from the ecologies to be restored" (Elias, Joshi, and Meinzen-Dick 2021, 7). Moreover, it has been noted that ecological restoration "must be understood as a wise investment with demonstrable economic returns in both monetary and nonmonetary terms" (Alexander et al. 2016, 1), with reforestation being pursued only when an economic rationale can be fostered along with the environmental potential. This global policy project does not acknowledge the interconnectedness of the ecological and human elements as being part of a shared planetary history; and this lack of acknowledgment opens up to the possibility that neoclassical economics and business as usual will be replicated and even reinforced in restoration projects (Büscher and Fletcher 2020). Nature (branded as "natural capital" and in terms of the "ecosystem services" it provides to humans) emerges in fact as a separate biophysical entity to be counted and later scaled up and commodified in the name of conservation; or stored and restored (often through technocratic means) when already on the verge of irreversible extinction.

The Ethiopian case may be particularly telling of large-scale ambitions, in turn translated into plans and investments—and not only with regard to restoration of nature. In fact the government, in response to the decline of natural forests, has engaged in establishing industrial plantations since the early 1970s. The phenomenon of state-led agrarian change, in collaboration with international corporate entities, has greatly contributed to shifting the crop focus from subsistence to commercial farming in several areas of the country. The detrimental impact of this transition has been especially felt by agro-pastoralists in Southern Ethiopia, in terms of environmental and livelihood strategies, loss of biodiversity, an increase in vulnerability to both food insecurity and food sovereignty, and an erosion of agency in pursuing local ways of producing food and choosing which crops/varieties to cultivate (Gebeyehu and Abbink 2022; Lund and Baudouin 2023). Against the backdrop of monocultural and other-directed landscapes, the plots of my ethnography feature plants and people engaged in intimate acts of interspecies care, tracing "the contours of modest forms of biocultural hope" (Kirksey, Shapiro, and Brodine 2013, 231).

In what follows I weave a counter-narrative around biodiversity conservation that pivots on one such case of unheard voices, omitted bodies of knowledge and experience in caring about natural resources, hard-to-commodify human-plant communities, and what I later call "quiet" conservation. Although relevant in mobilizing efforts and responses at the national scale, the approach of state policies to environmental management through mapping for restoration, and control over resources, is but one version of otherwise plural and deeper ecological histories of people and places.

This chapter starts by outlining the secret and highly interconnected life of an Ethiopian plant that for centuries has been symbiotically grown, sensuously appreciated, and fervently consumed on small farms, particularly in intercropped patches behind the house that have remained, by their very nature, sovereign precisely because they are unremarked or not countable. The thriving vitality of these agroecological niches has barely attracted public attention, research, or investment. Not coincidentally, in Ethiopia the dictates of the green revolution—which privilege productivity over social justice and food sovereignty, and calculation over taste and experience— have increasingly translated into processes of agrarian change and national nutritional guidelines that foster new modes of production and consumption through technological fixes, with no concerns about the ecological and sociopolitical implications of certain forms of agriculture or natural resource use and management. Nevertheless, the institutional logic has at times attempted to capture the economic potential of homegardens and bring such spaces of alternative agriculture and dissent under control through public display and measurement. I make a brief detour into these failed attempts, and then focus on the most recent initiative that the government of Ethiopia has undertaken toward (re)greening the country: the Green Legacy initiative, with its persistent rhetoric on restoration and planting to replenish what is lost, and the parallel, eloquent neglect of what, far from high-sounding pronouncements, is already strikingly green and still alive.

Biodiversity Is Enduring in Ethiopian Homegardens

Ensete (*Ensete ventricosum* [Welw.] Cheesman) is a long-lived banana-like perennial plant that is cultivated by smallholders throughout the Southern Highlands of Ethiopia (Rahmato 1995).[3] Ensete is unique to Ethiopia, which is in fact the only place where the plant has been domesticated and then cultivated starting from around ten thousand years ago (Brandt et al. 1997). Ensete has always been an orphan crop and has received far less attention from policymakers when compared to cereals and cash crops (Chivenge et al. 2015; Guinand and Lemessa 2001; Tsegaye and Struik 2002). This lack of attention is due not only to the nature of the plant in and of itself, but also to the fact that the plant generates and maintains a web of ecological interdependence and synergies with the surrounding environment that account for an ecology far more complex compared to monocultural (hypersimplified) systems.

In Ethiopia, the influence of the Millenium Development Goals toward high-tech and high-modernist scientific solutions for the "food global crisis," including the pursuit of food security through plants with universal application, has invigorated in national authorities a sense of urgency in shaping

citizens' diets by promoting approved forms of production, market participation, and dietary standards (Peveri 2021). In a country where measures of
yield have been consistently treated as the most important factor to farmers, research on the small plot sizes of homegardens, whose virtues are not
easily scalable, struggles to take root. The life cycle and seasonal rhythm of
perennials stand in sharp contrast to the annual-centric vision of the state.

Ensete provides food but also the opportunity to flexibly diversify production of different ensete byproducts (e.g., fibers, fodder, medicines, building
and wrapping material). The parts of the plant that are used for human consumption are the enlarged pseudostem and underground corm, which swell
over time with carbohydrates. Farmers transplant ensete individuals several
times during their life cycle; full maturity is reached after four to twelve
years—widely depending on variety, management, and climate. An elaborate
process is required to extract the starchy pulp from the pseudostem and
corm; several women from the community are recruited by a household
and come together to perform this highly cooperative and engaging task
(MacEntee et al. 2013). Women farmers pulverize the corm, squeeze the
leaf sheaths, and mix and then wrap the resulting pulp with several layers of
ensete leaves. This living material is stored safely behind the house to mature
in earth pits, with periodic remixing and kneading until it is considered to
be properly fermented (Fujimoto 2011). It is women who will later transform the fermented pulp into various foods—for the physical and spiritual
nourishment, as well as culinary enjoyment, of their families. The heavy,
bulky tubers of ensete provide a long-lasting underground store of calories
that is resistant to drought, pests, and disease outbreaks—which may, on
the contrary, have severe effects in monocultural stands. The ability to store
processed ensete pulp with little loss for long periods, lasting months or even
years, has provided households with a mechanism to modulate consumption
during food shortages.

The role of ensete in agronomic, nutritional, and ecological terms, however, cannot be grasped unless we place this perennial lynchpin where it
holistically belongs: the homegarden. In the places of my fieldwork, all ensete
inhabits small but highly populated plots around the homestead. Ensete cannot be read in isolation from multicrop and multispecies cooperatives, and
indeed it needs multiplicity as a mode of habitation to play a critical role at
both the plot and the pot levels. Ensete is not a standalone plant; it belongs
to a complex system whose components (that is, a mixture of perennials,

annual crops, and animals) all contribute to overcome some of its inherent weaknesses—including low vitamins and protein content, bacterial wilt, continual harvesting, and the need for manure to maintain vigorous growth (cf. Jacobsen et al. 2018). Ensete-cultivating groups have traditionally developed a high level of farming integration and have engaged in wide-ranging, and very subtle, combinations of livelihood activities to cope with structural bottlenecks. The plant is intensively intercropped to give nutritionally diverse and culinary enjoyable foods throughout the year; wild foods are gathered; and herbs, vegetables, and condiment crops complement grains. In juggling multiple crops, constantly changing plots and soil conditions, battling against new pathogens, and adjusting gardens to incorporate new plant species, gardening becomes a creative act that requires constant improvisation and innovation. Moreover, rural households are engaged in an integrated crop-livestock system.

One of the keys to the success of ensete-based systems is in fact maintaining livestock and hence access to manure as a source of soil fertility and thus of vitality of the ensete garden. Within this intricate web of species, in a forest-like microenvironment, ensete acts as a sort of "nurse crop" that hosts and shelters family members—be they human, animal, or vegetal beings (Board on Science and Technology et al. 1996, 179). In many ways the ensete homegarden can be considered as a (wo)man-made forest functioning as closed nutrient cycling.

In southwestern Ethiopia the homegarden cultivation of ensete represents a long-established example of an ecological system that thrives on diversity, integration of companion species, and complex traditional knowledge. According to available estimates, about 20 percent of the Ethiopian population, over 20 million people, depend on ensete either as a staple food or as a fallback crop in the event of seasonal or extreme food shortages (Borrell et al. 2019; Negash and Niehof 2004). Moreover, homegardens are typically not a supplementary production system in which only fruit and vegetables are grown to integrate staple crops grown in open and large fields. Instead, they are a principal livelihood system in which all forms of crops—including staple, cash, and supplementary crops—grow together. Found in smallholder farms of less than a hectare area, they are managed mostly by family labor (typically by women) with minimal to no use of chemicals or machinery. Their composition varies from area to area; however, their basic structure always comprises a combination of perennials as the stable components of the mixture. In Southern Ethiopia homegardens commonly take the form of perennial-crop-based systems, with ensete and coffee (both native species with life cycles of 8–12 and 24–30 years, respectively) acting as the backbone of this ingenious cultivation (see Abebe 2005; Kippie Kanshie 2002; Mellisse et al. 2018; Peveri 2020).

Perennials in the mixture keep these complex systems healthy and durable. In many ways, the perennial component acts as the secret ingredient of these mosaics of tiny patches, otherwise defined as "a patchwork of forest fragments in a matrix of agriculture" (Perfecto, Vandermeer, and Wright 2009, 4). In these mixtures, smallholders thoughtfully avoid relying solely on annual crops for immediate consumption, instead keeping perennials (e.g., ensete, coffee, fruit trees) to foster integrity and stability in the whole system. Perennials in fact occupy, feed, and benefit land for longer periods. Whenever they have been kept in the mixture, they are praised for providing beauty and diversity of species, for helping prevent a buildup of pests, and for benefiting wildlife (Altieri 2004, 37–38). Long-lived plants also bring their own conditions to humans: they last, take time to give back, and ask farmers for long-term planning, investment, care, and know-how. The reality of cultivating perennials is regulated by the life cycles, needs, and idiosyncrasies of the plants themselves (van der Veen 2014, 805). In one way, edible perennials lock farmers in through deep commitment, imply a long-term perspective, and shape specific forms of materiality and sociality. Within the family farm, gardening and horticulture articulate a rhythm complementing that of agriculture—one that runs slower and deeper.

[In gardening and horticulture] the focus tends to be on root and tuber crops, vegetables, herbs and fruits—cereals can also be grown in small garden plots—grown mixed together in the same garden, in contrast to large fields with one or two annual crops. Many of these plants thrive under intensive cultivation with regular monitoring and nurturing, using a strategy of "little and often," and garden plots are thus often located close to the home. The plants are, consequently, nearer and the plant-people relations closer, incorporating more handling of individual plants, such as planting, division and taking cuttings, as well as weeding, watering and feeding. . . . This closer familiarity between cultivator and plant may have given rise to greater experimentation and the development of new varieties, and may have generated a more prominent role for plants in people's ritual life, medicine, etc. (van der Veen 2014, 804–5)

Contrary to perennials, annual monocrops love being disturbed by humans through constant tillage by the plow, which breaks and erodes, and through injections of chemical fertilizers and pesticides. In the long term, this contributes to the negative environmental impact (caused by soil erosion, leakage of nutrients, and soil carbon loss) that is now documented worldwide (Crews and Rumsey 2017; Glover, Cox, and Reganold 2007). While better management of annuals would contribute to partially mitigating the phenomenon, a fundamental question remains around the potential of perennial polycultures to reverse the trend and tackle the root cause of the problem (cf. Batello et al. 2014).

Mainstream agricultural and conservation policies in Ethiopia have not yet addressed the question posed by existing and still thriving, but increasingly endangered, perennial polycultures. Because of their inherent socioecological complexity, along with the fact that "agroforestry is a complicated and knowledge intensive technology ill-suited to pre-fabricated farm-based packages" (Jerneck and Olsson 2013, 116), homegardens have remained a scientific mystery at the global level.[4] In a nutshell, if these multispecies systems have not received research and policy attention, "this is primarily because they do not fit into the single-species model of agricultural development paradigms" (Nair 2017, 4). In other but complementary terms, ensete's air of secrecy could be read as a strategy pursued by those who lovingly tend the plant to preserve the rich Indigenous knowledge that has flourished around it, and to silently foster their own food security and sovereignty by resisting extractive practices through localized responses. This seems in line with the classical argument of James C. Scott (1998), who demonstrates that certain forms of autonomy come from invisibility/illegibility, suggesting a possible reading of homegardens as self-contained spaces of sovereignty.[5] In fact, whereas many annual crops require external inputs to be productive and thus provide incentives for private-sector involvement, perennial crops generally require less or no planting material or other inputs from outside. This way, farmers keep a part of the landscape beautiful and gastronomically enjoyable on their own terms. Edible gardens are clearly intended to function outside the centralized system of control, helping families to develop agronomic and gastropolitical habits of their own choosing.

With this unruly complexity in mind—embodied in tall and evergreen ensete, and in the flourishing and intricate vegetation of ensete-growing areas
where, on the margins of market economies, a single cultivated plot typically
resembles a botanical garden, filled with many species of plants—we now
move from the ground up to the institutional level of guidelines, campaigns,
and (re)greening initiatives.

Mapping the Intricacy of the Ensete Garden

The case of a root tuber crop, half hidden, half in plain sight, and literally
buried deep in homegardens, is an intriguing entry point into the dilemma of
how embodiment can or cannot come to terms with inscription. At different
points in time an inherent tension has surfaced between ensete-growing
communities who bask in the secrecy of their gardens and the institutional
logic that aims at trimming this intricate entity down through public display
and measurement. Here I present two snapshots of the ensete story and
highlight how these human-plant underground communities have developed over time a habit of deception as a way of coping with institutional or
corporate attempts at penetrating into the homegarden.

Before leafing through the snapshots it must be noted how, among ensete-
cultivating groups, farms are usually divided into gardens and fields (Abebe
2005; Peveri 2020). Grain crops, like anything grown for cash, strongly identify with men. The field is the sector of the small farm where rural households risk experimenting with newly introduced, high-yielding varieties in
an attempt to better their socioeconomic position—by producing, at least
in part, for the market. Nonetheless, a small farm may take risks in the field
only if another sector (the garden) guarantees the survival of the whole organism. The household splits into a sector that takes risks and is legible to
the state and a sector that monitors subsistence and is not legible. If agriculture may demand legitimization, horticulture seems less so. The illegibility
of homegardens in southwestern Ethiopia is eloquently revealed by the fact
that ensete has been traditionally defined in historical and anthropological
literature—in quite sterile terms—as the "tree against hunger" (Brandt et al.
1997). This official definition is strongly at odds with the symbiotic and sensuous dialogue the plant enters into with its human companions.

The farmers I worked with think of each exemplar as having individual
appearance, sensibility, charisma, and taste, as an epiphany of beauty and

prosperity, and even believe that the plant can feel pain. Its cultivation requires continuous acts of care that leave traces in the bodies of its (women) caretakers and in the whole socioecological fabric. The nuances of color and personality in the plot are reflected in nuances in textures and flavors when the starchy material reaches the pot. When registered in institutional narratives, this perceptual and affective engagement is conflated with the concept of mere survival. The tacit yet axiomatic idea that people in need require food, any food, for adequate nutrition and sustenance is not uncommon in either humanitarian projects or national policy initiatives. According to this logic, subsistence or traditional farmers are treated not as partners but as silent recipients, or "taste-less subjects," whose right of food choice is therefore framed within what could be defined as "the politics of necessity."

The "taste of necessity" is built on an erasure of sensory experiences and on the general assumption that everything edible should be considered food, with people focusing on actual starvation rather than, as is indeed the case, on cultural (and socioecological) survival (Bourdieu 1984; Trapp 2016). Yet, Hadiyya farmers and cooks represent an example of people who spend a considerable amount of effort and time in growing, harvesting, processing, and cooking foods that would appear simple, nutritionally poor, or ugly to an outsider, but that they nevertheless perceive as being socioculturally and affectively important. For them, taste and value are both driving factors in determining what should be cultivated, eaten, saved, and maintained.

The first scene of the ensete history is set at the end of the nineteenth century, when Emperor Menelik II embarked on a campaign of expanding his rule from the central highlands to the peripheries of Ethiopia. In the southwest, the most significant cultural impact was felt in the changeover from hoe cultivation to plow agriculture (Orent 1979). This transition involved a shift from perennial to annual crops, and from root crops to cereals, because of their symbolic value but especially because, unlike root crops, cereals could easily be counted and stored. This shift opened up opportunities for ensete-cultivating groups to widen and diversify their rural agricultural ventures. They never completely left ensete behind, however, and continued to practice their love for it as an act of secrecy—secrecy in full view but tight enough to avoid being squeezed into enumerability.

According to the seminal work of W. A. Schack on the Gurage group, neighbors of the Hadiyya, the eating behavior of local communities and their attachment to ensete have been molded by persistent political conditions, among them "the political expansion of the powerful Amhara kingdoms from the seventeenth century onwards." The Gurage developed a vision of the world "as being threatening, hostile, and fraught with insecurity" (Shack 1971, 40). A plant that can be stored underground and consumed sparingly was at that time and has continued to be the most comforting antidote to fear and suspicion from external threats.

> The storage of large quantities of ensete in deep earth-pits, and the deliberate hiding of the exact locations of the storage pits (by covering over the openings with garden refuse), are cultural practices which have been exacerbated by the historical dangers of destruction. For in the past, after the most devastating plundering and raiding of villages for slaves and cattle, the ensete-food buried deep in the earth eluded the pillage of hostile raiders, to be recovered later by survivors once the threat of danger had passed. (Shack 1971, 40)

Throughout history, marginalized groups have communicated using languages that were intentionally opaque to outsiders; in the case of ensete growers, profound truths have often lurked in socioecological refuges inhabited by a dense understory and by deep-rooted species.

The second snapshot is set around 2012 precisely in the districts where I was conducting my fieldwork. The consultant of a big Canadian company (Potash Corporation of Saskatchewan Inc.) arrived to assess the yield potential of ensete with the long-term perspective of lifting farmers from the great weight of this plant and providing them with an "easy" crop that would instead boost productivity and ensure them a brighter future—hybrid maize.[6] In the end, the consultant was unable to standardize the calculation of plants per hectare because of the very nature of uncountable ensete, which can be held in reserve for five to ten years, and which farmers have the option to use at any time after maturity. No numerical formula could be applied to ensete for estimating its production and leveling out fluctuations in yield. Moreover, ensete is vegetatively propagated by using suckers, not seeds; thus, like other root tuber crops, it lies beyond the proprietary control of agricultural capital, being virtually impossible to wrest control of the plant's reproduction from the farmer. In other words, the ensete-based polycultural system proved to be nonscalable (cf. Peveri 2020, 184–86).

Through the trajectory of ensete, another broader trajectory can be read: that of composite forms of smallholder farming toward which national and global development policies have shown an age-old and still lingering lack of research and public investment (Borrell et al. 2019). The history of mosaic landscapes in southwestern Ethiopia is punctuated by official attempts at governing people and land since imperial times and, once those head-on attempts failed, by the adoption of agricultural policies whose indirect reverberations in the long term would likely be to crumble their structural complexity into simplified and more manageable units. Not coincidentally, in Ethiopia, traditional polycultural systems—of which edible perennials are a major component—are rapidly replaced by hybrid maize and other forms of monoculture (like tea, pepper, and eucalyptus).[7] In the areas where I conducted my ethnography, the decline in the share of these perennial components during the past decade and their replacement with annual crops (particularly hybrid maize) has been of great concern, especially for how this shift could reduce the ecological benefits and long-term sustainability derived from these complex systems.

The commercialization of agriculture is a major factor in determining the loss of food crops, along with policymakers and researchers' persistent neglect of Indigenous plant species. In the capitalist world ecology, "the denial of complexity is the common denominator to every sacrifice zone. Arranging nature in the interests of capital requires a mass simplification: the reduction of all life into the categories of resource or waste" (Farrier 2019, 52; see also Sheridan 2016). A neoliberal approach to natural resources may well represent a key to contextualizing the persistent institutional inclination of Ethiopian elites toward simplified ecologies, and their studied unseeing of dense entanglements.

And yet, what the story of ensete articulates with even more clarity is the extreme resilience of homegardens, and their capacity to bloom over time despite extraction, neglect, or dispossession. Gardens are replete with unruly potential while resisting appropriation because of the apparent chaos, and the menacing alterity of multiplicity, that reigns at the edges of official forms of agriculture and greenery. Gardens' resilience and vitality lie in botanical combination. The notion of beauty for farmers who live in close connection to ensete lies in species diversity and complementarity, not in the high-risk concept of "one best variety."

Explorations into the intricate relations taking place in a "simple" home-garden have become intense in the last twenty years, with development scholars, ethnobotanists, and ecologists increasingly pointing to homegardens as important sites for the maintenance of plant biodiversity and alternately as "repositories," "sanctuaries," or "medicine cabinets" (Eyzaguirre and Linares 2004; Kumar and Nair 2006). Homegardens are indeed experimental plots where biocultural diversity is preserved and creative solutions devised to resist the unsustainable shift to monocultural thinking.

What is relevant in conceptual terms is that this work on biodiversity conservation in the plot—of saving and sharing tools, seed, shoots, skills, and produce, and making them proliferate—can be considered under the umbrella of widespread "actually existing sustainabilities"; that is, of "what appear to be longstanding sustainable practices that the practitioners themselves, and the research and policy community, have not recognised or valorised in those terms" (Smith and Jehlička 2013, 148). The framework of "quiet sustainability" coined by Smith and Jehlička (2013) on the understudied subject of food self-provisioning in postsocialist Central and Eastern Europe aims at capturing the blind spot of sustainability that is practiced at the level of the household, through informal (family and friendship) networks, and that has rarely entered formal planning or research agendas. Through this concept, the value of resilient gardens for self-provisioning, as well as for pleasure and commensality, is highlighted. Its explicative power extends well beyond the context discussed by the authors to include everyday practices that are common in the Global South.

> Quiet sustainability is defined by practices that result in beneficial environmental or social outcomes, that do not relate directly or indirectly to market transactions, and that are not represented by the practitioners as relating directly to environmental or sustainability goals. Cultures of sharing, repairing, gifting and bartering characterise quiet sustainability. Everyday practices that have low environmental impacts, but that have not been pursued for that reason, are also features of the concept. (Smith and Jehlička 2013, 155–56)

The biodiversity work carried out in Ethiopian homegardens is equally quiet but full of purpose and will; it opposes in many ways the leveling logic of "one" market and "one" ecology, and mimics the complexity of nature rather than the simplification of politically imposed uniformity.

The illegible and partially ungovernable nature of multifunctional mosaics challenges modernity's desire to force reality into quantitative molds and contrasts numerical approaches to landscapes and plants. Not coincidentally, there is no trace of this intricate vegetal excess in the Green Legacy discourse of the Ethiopian government. The sense of natures as becoming—as sites of energy and transformation—freezes into scores to be reached, billions of trees to be planted, and the undercurrents of productivity, efficiency, and control as a corollary to be gained from restoration. The ethics-in-place (Alaimo 2016, 30) found in the ensete garden makes way for ecosystem restoration as a technical project run by (mostly male and middle-class) humans, as well as to a discursive regime that fails to include nonhumans and other living ecological stories.

Counting, Restoring, Replanting: The Green Legacy Initiative

On June 29, 2020, a popular Oromo singer, Hachalu Hundessa, was killed on the outskirts of Addis Ababa. A burst of intercommunal clashes and unrest in both the capital city and the Oromia zone followed and, at the time of writing, in early August 2020, had claimed some 160 lives. This chain reaction was triggered by several factors not in the scope of this chapter to discuss, but among which commentators list the deep historical roots of Oromo nationalism and the discontent with Prime Minister Abiy Ahmed Ali's policies.[8] This was the worst bout of protest, and harshest revival of ethnic identity politics, since the PM came to power in April 2018.

The day after the killing, the government imposed a curfew and a blanket internet ban to defuse (or so they claimed) the escalating divisions, factionalism, and ethnic-based violence in an already combustible political situation; most Ethiopians were cut off from the internet connection for about three weeks, and opposition leaders were arrested. Some critics have pointed to what they consider a heavy-handed state response in a traditionally ethnicized political system; according to this reading, these recent events would in turn anticipate and mirror the progressive consolidation of power through law and order, pursued by the government to quell the resurgence of competing and divisive ethnonationalisms. In such a polarized and emotionally charged political climate, forms of mobilization around the ideal of a strong and centralized state started once again being fed into the public opinion with greater intensity, and on various media outlets. This call for unity included a pounding rhetoric (and call to citizens for support) around the mega-project of the Grand Ethiopian Renaissance Dam through the slogan and hashtag #ItsMyDam, as well as the frenetic growth of expensive infrastructural modernization in the capital city—such as the controversial Unity Park, a newly opened tourist attraction at the heart of Addis Ababa, built at a cost of 170 million dollars (paid by the United Arab Emirates, a close ally), which also features a botanical garden.[9]

What is striking in the convergence of these political trends toward grandeur and the accomplishment of national imperatives, however, is that the official Twitter account of the prime minister has been recently filled neither by the Renaissance Dam nor by the infrastructural renaissance of the capital city, but by the (apparently simple) act of planting trees.[10] In the days immediately preceding the death of Hachalu Hundessa, photographs of federal and state ministers kneeling down to dig and plant seedlings populated the account of Abiy Ahmed. Encouragements followed, especially directed toward the youth, to leave a green mark by engaging in volunteerism in service of their local communities, as well as a special call to all Ethiopian creatives—filmmakers, graphic artists, photographers, storytellers, and cartoonists—to submit videos celebrating the #PMGreenLegacyChallenge that could "inspire and motivate communities to plant their print this year." After the three-week gap in communication, the daily tweeting activity resumed. In one of the tweets Abiy, in a summer military uniform, is shoveling soil; the caption recites: "What we aim to achieve, we can deliver on together. Plant peace. Plant a lasting legacy!"

The scheme of cleaning and "greening Ethiopia one seedling at a time" through a National Green Development Programme was launched in May 2019 as part of the Bonn Challenge, with the ambitious pledge to plant 4 billion trees across the nation and restore 15 million hectares of degraded forests and landscapes by 2030 (which means that forty trees per person should be planted).[11] In the official words of its promoters, the afforestation initiative is meant to mitigate the effects of fast-paced climate change (including flooding, drought, and food insecurity), combat environmental degradation, and transform Ethiopia into a carbon-neutral middle-income country by 2025. The initiative, which has been highly praised by international agencies since its start, is rhetorically centered on numbers whose accountability remains uncertain. On July 29, 2019, the minister of innovation and technology was quoted saying that more than 353 million trees were planted in twelve hours, breaking the world record held by India since 2017 for the most trees planted in one day, which stood at 50 million trees.[12] Ethiopia is now considered a model country from many quarters.

Only spare information is available about long-term seedling survival after each performance, however, and no data are available on what those different "eco-friendly seedlings" actually are. Moreover, no explicit mention of existing agroforestry or traditionally biodiverse farming systems is made on the official platforms that otherwise claim how newly planted trees are expected to cover millions of hectares of land with forest. But what kind of forests does the Ethiopian government plan to create? As paradoxical as it might appear, the whole notion of what species should be used in restoration has received little attention so far, and data are currently missing in this regard. In the uncertainty of what is being planted, it cannot be excluded that much of this growth would be in monoculture plantations. And indeed, an assessment published in the journal *Nature* comparing these large-scale operations in different countries shows that 45 percent of promised new forests will be monoculture timber plantations of fast-growing trees, usually destined for harvesting in double-quick time to make pulp for paper (Lewis et al. 2019).

Critical remarks are mounting in the scientific literature as well as in opinion news on initiatives of this kind, which are on the rise in Ethiopia as in many other (developed and developing) countries—and which some ecologists consider nothing more than a "fairy tale," or misleading at best (Pearce 2019). The main counterargument is that planting more trees may indeed create new environmental problems, including the threat they would pose to some of the country's ecosystems. An online article in *Quartz Africa* argues that "for the initiative to work, trees planted in the country's different ecological environments need to be tailor made for their location. If the right trees are not planted in the environments for which they are a fit, the 'Green Legacy' might be doing more harm than good" (Hadero 2019). This is especially true in a topography that varies as widely as Ethiopia's—ranging from the depths of the Rift Valley, 126 meters below sea level, to the elevations of the highlands at 4,620 meters.

For example, not all ecosystems would benefit from more trees and a thicker canopy; in tropical grassy biomes, planting trees might push other plant and animal species away in search of open and sunny locations, therefore contributing to a loss rather than a gain in biodiversity. Projects fail because they choose the wrong trees, use too few species, or are not managed for the long term. In brief, most projects do not think ecosystemically. If unity and uniformity may well be nurtured at the political level to achieve national reconciliation, there is no one-size-fits-all governance, nor centralized global policies, that can effectively and sustainably respond to the call of preserving or increasing biodiversity.

A cold sort of engineering approach to designing and restructuring ecology (that is, a strategic policing of space) is not new in Ethiopian history. Past waves of afforestation efforts have unfolded through the introduction of damaging nonnative species; for example, eucalyptus, which, due to its quick growth and inability to recycle rainfall or capture carbon dioxide from the air, has proven to be destructive on land by causing soil acidity and destroying nearby groundcover (Lemenih and Kassa 2014). Careful monitoring of the impacts and failures of past afforestation programs has not informed the new greening policies, which keep marching in the name of mass plantings to meet a national quota. At this point, beneath the ecological and possibly genuine concerns about biodiversity loss and land degradation, we see how politically charged the tweeting and broadcasting activity around the Green Legacy initiative is. It is no coincidence that, at a time when the government faces ongoing struggles in governing the multiethnic country, a conservationist posture with regard to natural resources has emerged as an opportunity to cement the state-building project.

The attempt at the roots of the Ethiopian campaign is to re-create "natural forests" wherein existing forest lands are increasingly endangered and shrinking, and agroforestry systems have been historically neglected in agricultural policies. A recently released report by a coalition of German and African civil society groups, which underscores the failures of the Alliance for a Green Revolution in Africa and similar models of agriculture, also highlights the contradictory approaches of Ethiopian institutions.[13] While some agroecological farming practices have in fact been integrated into policy, Ethiopia has concomitantly known a state-led aggressive (and often coercive) expansion of artificial fertilizers and hybrid seeds; has seen the proliferation of improved wheat varieties that are short lived and vulnerable to rust and other diseases; and has recently opened up to genetically modified crops (Hunduma 2020).[14] Can the faith in techno-scientific solutions that has always been rampant within government agricultural institutions at least partially explain the condescension toward smallholder farming? And, more specifically, toward homegarden agroforestry practices that do not require the use of modern technologies but rather the optimal use of local resources and natural processes?

According to the above-mentioned report, agroforestry would not offer as many ecological benefits as those provided by natural forests, and yet agroforestry typically holds six times more carbon than monoculture plantations; moreover, in heavily populated areas, agroforestry provides food and a wide variety of other goods along with long-term ecological and economic resilience. Given that natural forests are long maturing and would therefore badly fit into a global agenda of visible results in short time frames, one viable and realistic option would be for policymakers to commit to reforesting primarily through agroforestry. This should be done in close collaboration with civil society, marginalized groups, and particularly smallholders, whose interests as stewards and caretakers of the land need to be better integrated with those who have the money and loud voice to support restoration (Cernansky 2018). In fact, as pointed out by Perfecto, Vandermeer, and Wright, "agro-ecosystems are important components of the natural world, intricate to biodiversity conservation. Consequently, their thoughtful management should be part of both a rational production system and a worldwide plan for biodiversity conservation" (2009, 3). Evidence is growing that, "in most cases, some form of multispecies, or at least genetically diverse, plantings is needed to restore ecosystem function" (Alexander et al. 2016, 3). Yet, despite evidence of the role that mosaic landscapes can play, Green Legacy discourse makes no mention of mosaic restoration, in which forests are embedded into agricultural landscapes, nor shows any interest in studying the role of native species and their durability and robustness. While planting trees is an act of high symbolic value that easily marshals consensus, keeping trees represents indeed the challenge in answering the pressing question posed by Elias, Joshi, and Meinzen-Dick (2021): "restoration of what, by whom, and for whom?"

There are growing concerns worldwide—against the backdrop of predominant economic models that favor short-term gains over long-term sustainability—that the reforestation agenda may become a green cover for further assaults on ecosystems (Alexander et al. 2016, 2; Igoe and Brockington 2016). Certain scholars invite a constant state of vigilance over insidious forms of what they define as "the rampant expansion of 'bourgeois environmentalism'" (Elias, Joshi, and Meinzen-Dick 2021, 10). One of the mythological features of the global discursive regime around conservation is the existence of "natural communities" to be restored; but have those forms of "intact" nature—that environmental do-gooders now claim should be walled, gated, or forted up—ever existed? Indeed, much conservation work concentrates on protecting isolated and scattered fragments of natural vegetation (erroneously thought of as pristine) and ignores the more complex (mostly agricultural) matrix and myriad connections to human activity in which they exist (Perfecto, Vandermeer, and Wright 2009, 7–8; see also Angus 2017, 125–30).

Moreover, what are the implications of these greening projects in terms of gender, class, ethnicity, and other intersecting axes of difference when policymakers determine which trees should be planted, why, and how? The risks inherent in neoliberal environmentalism, especially in seeding fast-growing tree species to reach standards of quick greenery, is the potential erasure of landscapes that are outside grand schemes for the preservation or commodification of nature. In those small places the Green Legacy has taken a long time to thrive; they are feminine and delicate by nature; they are landscapes in which quiet, unspectacular conservation is the norm to ensure economic, human, and multispecies well-being.

The restoration framework proposed and implemented in the Ethiopian Green Legacy initiative, as inspired and compelling as it appears through its media campaign, remains oblivious to the structural complexity of multispecies systems. Accordingly, the initiative does not even take into consideration the rehabilitation of traditional agroforestry systems—through the recognition, protection, and nurturing of "practices that are sustainable in outcome, but don't seek or claim to be, and that happen apart from, sometimes in spite of, the economic sphere" (Smith and Jehlička 2013, 155). As with the ensete homegarden, large sections of what for local communities means staying with nature and doing biodiversity are neglected, forms wherein "there is not a fulfilment of environmental obligations, an attempt to achieve 'resilience,' or a response to limits, but the daily practice of a satisfying life" (Smith and Jehlička 2013, 156).

Such erasures become apparent when reviewing the visual materials of the Green Legacy campaign. A man in an elegant business suit and tie or in military fashion is planting trees; the few women portrayed look like civil servants, not farmers. If efforts to promote pluralism, inclusivity, and diversity are to be found, they are directed toward rendering the many souls of an urban middle class. These are visually striking examples of how gender and class divisions are projected into tree planting for the international audience. From this angle of vision, what is privileged is not the small scale, nor the human body of those who live in close association with plants and relate to the landscape as if they belonged. Least of all, do such initiatives succeed in capturing the standpoint of the plants themselves. And thus, indeed: restoration of what, by whom, for whom, and for how long?

Conclusion: Quiet Conservation in Times of Environmental Grief

There is no easy way to reconcile biodiversity rooted in place and for a good life, as it is found in the ensete multispecies garden, with biodiversity in display cases that characterizes most of the globalized conservation debate. The more the rarefaction of bodies and humoral encounters came to mark my daily life, the more my memories of those gardens turned into searing physical pain—into something imagined, felt, and wanted. They were indeed refuges of extensive pleasure, of aesthetics and politics, of abundant sensuality and playful experimentation. This bodily recollection of the plant, places, and people—for which I still feel "crazy love" in the sense indicated by D. Bird Rose (2013)—is opening up to less corporeal reflections on how to build a new paradigm of conservation one piece at a time.

In this regard, the pandemic acted as a tremendous catalyst for considering how reuse can be a viable mechanism in times of crisis, especially since the effects of this crisis will stay with us. On the one side, (re)planting anew is good for propaganda, aligns with confident but unrealistic predictions, and speaks of aspirations to redeem, tame, and remake nature; on the other, the act of restoring and resuscitating assumes decaying or half-dead bodies, which would therefore be tagged, counted, described, and possibly removed to be held in climate-controlled environments for safekeeping (cf. Bird Rose 2013, 14; Harrison 2017).

In between lies the quiet and revolutionary option of maintaining and nurturing, as well as learning from, already existing ethical forms of relating to the environment and to ourselves as a species. This third option is rarely considered in "greening" plans, as it would require delineating an ethics of care and custodianship in the nooks and crannies of the productivist approach, as well as, especially in contexts that have a long history of elite capture of benefits, actively working toward the delegitimization of conservation projects. More generally, this third, bolder option would call for an emancipatory politics that progressively transitions from controlling (visions of planetary management) to caring; from singularity and scalability to meaningful diversity; and from growth per se to collaborative flourishing (Scoones and Stirling 2020).

The local small-scale history I have recounted of the ensete communities of my fieldwork sheds some light on their experience of ecological pressures and change over time. The long trajectory of ensete caretakers suggests there is no safe path from the sensuous to institutional recognition and support unless the secret of joy and beauty is, at least in part, protected against other interests. For those human-plant communities, exclusion from the national banquet has proved several times to be a blessing in disguise. While scholars and activists, including myself, struggle to preserve, rescue, conserve, and document pockets of biocultural diversity in the face of massive environmental suffering and neoliberal forms of extraction, certain people, much like the perennial plants they care for, enter into periods of dormancy, disappear off the institutional grid, and fall into temporary oblivion.

Despite their geographic fixity, those humans and plants find "possible lines of flight" to escape the enumerative enthusiasm of state and corporate agendas. Then, slowly, they come back to full life. This intermittent movement is suggestive of how much we still have to learn from plants about the arts of resistance and a politics of hope—especially perennial ones with deeper roots that have enormous capacity to self-heal, to break off and start up again, not as coldly stored heritage but as pulsating and edible biodiversity.

It remains nonetheless a challenging task to find ways to preserve the sense of nature as becoming, make space for plant memory and culture, and infiltrate embodied orientations to (and lived interpretations of) the environment into the global (re)greening agenda. The first step toward an ecocentric model—one that is played in polyphonic relationships—might be to grow ourselves skeptical of numbers whose accountability remains dubious and instead look and ask for audacious vision. In that vision, even an inconspicuous homegarden challenges the urgent present of capitalism by infiltrating the deep time of the Earth's body into discussions of ecology. In that vision, informed action is required by competent and compassionate leadership that orients itself squarely toward reckoning with history and is willing to offer a less comforting kind of hope—not short-term but generative hope, in the form of what has been called "convivial conservation" (Büscher and Fletcher 2020), that can also be renamed as caring ecologies within or beneath neoliberal frameworks.

Notes

1. Similar reflections on armchairs and the impossibility of fieldwork (and, to a certain extent, of thought) are shared by poet and anthropologist Atreyee Majumder in a blog article, "On Armchairs and Catastrophe," *Cafe Dissensus Everyday*, May 16, 2020, https://cafedissensusblog.com/2020/05/16/on-armchairs-and-catastrophes/.

2. *Through the Looking Glass* is the second in the trilogy of Alice's adventures created by Lewis Carroll, when she consciously decides to reembark on her journey of discovery by visiting the world beyond the looking glass. The concluding words of the novel are, "Life what is but a dream?"

3. Perennial plants are those whose life cycle lasts more than two years (from seed to bloom to seed). They consist of different types, including perennial grasses, shrubs, trees, grains, and legumes. They emerge from dormancy at the end of the winter, or a dry season. Many of them stand in place for several years; some are capable of surviving for even a century or more. From an agricultural perspective, what makes perennials unique is that they require no year-to-year reseeding or replanting. They are often defined in opposition to annuals, which are planted from seed, grow to maturity, produce seed or fruit, and then die—all in one year. It is worth noting, however, that the life cycle of ensete significantly differs from that of perennial fruit crops; ensete has in fact a long life but is harvested only once before flowering (unlike banana) to avoid reallocation of resources from edible storage organs to the inedible inflorescence.

4. In fact, despite having "provided sustenance to millions of farmers, and prosperity to some, around the world for centuries"; and despite having "fascinated scientists for quite some time," "the extent of scientific studies on these systems has been disproportionately lower than what their economic value, ecological benefits, or sociocultural importance would warrant" (Nair 2001, 240). See also Kumar and Nair 2006, 137; and Borrell et al. 2019.

5. The concept of legibility, developed by J. C. Scott in *Seeing Like a State* (1998), refers to a state's attempt to make society legible, to arrange the population in ways that simplify the classic state functions of taxation, conscription, and prevention of rebellion. This has been typically achieved by geographic concentration of the population and the use of high-value forms of cultivation, to minimize the cost of governing the area as well as the transaction costs required to appropriate labor and produce. Grain crops have long been integrated into the modern market system and made "legible" to the state. This means that a percentage of crop production is extracted from farmers in the form of rents, taxes, costs of milling, transporting and

irrigation, and market intermediaries. In many developing countries this extraction has increased over the past decades through green revolution development projects, which have introduced productive but costly packages of high-yielding varieties of seeds, fertilizers, pesticides, and herbicides.

6. For a historical account of the "corn madness" that has spread over decades in the form of hybrids, triggering several waves of agroecological change in the country, see McCann 2007, 190–91.

7. The hybrid maize breeding program has a history of over thirty years in Ethiopia, in particular, the release of the BH660 and BH540 varieties to lay a strong foundation for this kind of technology, along with intensified efforts to extend and popularize it undertaken by Sasakawa Global 2000 (SG2000) in the mid-1990s (McCann 2007, 186). The SG2000 program is an agricultural initiative of two nongovernmental organizations, namely, the Sasakawa Africa Association (SSA), whose president was Norman Borlaug, and the Global 2000 program of the Carter Center, whose chairman was former U.S. president Jimmy Carter.

8. The Oromo are the single largest ethnic group in Ethiopia, making up roughly 40 million of the country's population, and the most internally diverse of Ethiopia's nations and nationalities. Part of the Oromo narrative is that this sheer demographic size has not translated into political and economic power, and that their history has been one of marginalization and exploitation from the ruling elites.

9. The significance of unity and cohesion inherent in the park is made apparent by its location—the Grand Palace, a compound that has been a seat of state for more than a century. The palace was initially constructed in 1887 as part of the founding of the city of Addis Ababa by then Emperor Menelik II and his wife, Empress Taitu.

10. Abiy Ahmed Ali (@AbiyAhmedAli), Twitter, accessed August 5, 2020, https://twitter.com/AbiyAhmedAli.

11. "Greening Ethiopia, One Seedling at a Time," Embassy of the Federal Democratic Republic of Ethiopia, June 17, 2019, https://www.ethioembassy.org.uk/greening-ethiopia-one-seedling-at-a-time/; "Green Legacy," Office of the Prime Minister, Ethiopia, accessed August 5, 2020, https://pmo.gov.et/greenlegacy/. The Bonn challenge is an international agreement launched in 2011 by the government of Germany and the International Union for Conservation of Nature (IUCN) to add 1.35 million square miles of forests (an area slightly larger than India) to the planet's land surface by 2030. It has received ample commitment from countries around the world, but the strategies are not always backed by evidence, and measures of success are still being defined. "About the Challenge," Bonn Challenge, accessed May 31, 2024, https://www.bonnchallenge.org/about.

12. "Spotlight on Ethopia's Tree-Planting Programme," UN Environment Programme, December 27, 2019, https://www.unenvironment.org/news-and-stories/story/spotlight-ethiopias-tree-planting-programme.

13. "False Promises: The Alliance for a Green Revolution in Africa (AGRA)," *Rosa Luxemburg Stiftung*, audio, 26:38, https://www.rosalux.de/en/publication/id/42635/false-promises-the-alliance-for-a-green-revolution-in-africa-agra. See also Hunduma 2020.

14. See also the Food and Land Use Coalition's newly released *Action Agenda for a New Food and Land Use Economy in Ethiopia*, January 2020, https://www.foodand landusecoalition.org/wp-content/uploads/2019/08/FOLU-Action-Agenda-Ethiopia _WEB.pdf.

References

Abebe, Tesfaye. 2005. *Diversity in Homegarden Agroforestry Systems of Southern Ethiopia*. Tropical Resource Management Papers 59. Wageningen: Wageningen University and Research Centre.

Aisher, Alex, and Vinita Damodaran. 2016. "Introduction: Human-Nature Interactions Through a Multispecies Lens." *Conservation and Society* 14 (4): 293–304.

Alaimo, Stacy. 2016. *Exposed: Environmental Politics and Pleasures in Posthuman Times*. Minneapolis: University of Minnesota Press.

Alexander, Sasha, James Aronson, Oliver Whaley, and David Lamb. 2016. "The Relationship between Ecological Restoration and the Ecosystem Services Concept." *Ecology and Society* 21 (1): art. 34. http://dx.doi.org/10.5751/ES-08288-210134.

Altieri, Miguel A. 2004. "Linking Ecologists and Traditional Farmers in the Search for Sustainable Agriculture." *Frontiers in Ecology and the Environment* 2 (1): 35–42.

Angus, Ian. 2017. *A Redder Shade of Green: Intersections of Science and Socialism*. New York: Monthly Review.

Archambault, Julie Soleil. 2016. "Taking Love Seriously in Human-Plant Relations in Mozambique: Toward an Anthropology of Affective Encounters." *Cultural Anthropology* 31 (2): 244–71.

Batello, Caterina, Len Wade, Stan Cox, Norberto Pogna, Alessandro Bozzini, and John Choptiany, eds. 2014. *Perennial Crops for Food Security: Proceedings of the FAO Expert Workshop*. Rome: Food and Agriculture Organization of the United Nations.

Bird Rose, Deborah. 2013. "In the Shadow of All This Death." In *Animal Death*, edited by Jay Johnston and Fiona Probyn-Rapsey, 1–20. Sydney: Sydney University Press.

Board on Science and Technology for International Development, Office of International Affairs, and National Research Council. 1996. *Lost Crops of Africa*. Vol. 2, *Vegetables*. Washington, D.C.: National Academy Press.

Borrell, James S., Manosh K. Biswas, Mark Goodwin, Guy Blomme, Trude Schwarzacher, J. S. Pat Heslop-Harrison, Abebe M. Wendawek, Admas Berhanu, Simon Kallow, Steven Janssens, Ermias L. Molla, Aaron P. Davis, Feleke Woldeyes, Kathy Willis, Sebsebe Demissew, and Paul Wilkin. 2019. "Enset in Ethiopia: A Poorly Characterized but Resilient Starch Staple." *Annals of Botany* 123 (5): 747–66.

Bourdieu, Pierre. 1984. *Distinction: A Social Critique of the Judgment of Taste*. Cambridge, Mass.: Harvard University Press.

Brandt, Steven A., Anita Spring, Clifton Hiebsch, J. Terrence McCabe, Endale Tabogie, Mulugeta Diro, Gizachew Wolde-Michael, Gebre Yntiso, Masayoshi Shigeta, and Shiferaw Tesfaye. 1997. *The "Tree Against Hunger": Enset-Based Agricultural*

Systems in Ethiopia. Washington, D.C.: American Association for the Advancement of Science.

Büscher, Bram, and Robert Fletcher. 2020. *The Conservation Revolution: Radical Ideas for Saving Nature Beyond the Anthropocene.* London: Verso.

Cernansky, Rachel. 2018. "How to Plant a Trillion Trees." *Nature News,* August 29. https://www.nature.com/articles/d41586-018-06031-x.

Chivenge, Pauline, Tafadzwanashe Mabhaudhi, Albert T. Modi, and Paramu Mafongoya. 2015. "The Potential Role of Neglected and Underutilised Crop Species as Future Crops under Water Scarce Conditions in Sub-Saharan Africa." *International Journal of Environmental Research and Public Health* 12 (6): 5685–5711.

Crews, Timothy E., and Brian E. Rumsey. 2017. "What Agriculture Can Learn from Native Ecosystems in Building Soil Organic Matter: A Review." *Sustainability* 9 (4): 915. https://doi.org/10.3390/su9040578.

Elias, Marlène, Deepa Joshi, and Ruth Meinzen-Dick. 2021. "Restoration for Whom, by Whom? Exploring the Socio-political Dimensions of Restoration." *Ecological Restoration* 39 (1–2): 3–15.

Eyzaguirre, Pablo, and Olga Linares, eds. 2004. *Home Gardens and Agrobiodiversity.* Washington, D.C.: Smithsonian Books.

Farrier, David. 2019. *Anthropocene Poetics: Deep Time, Sacrifice Zones, and Extinction.* Minneapolis: University of Minnesota Press.

Fujimoto, Takeshi. 2011. "The Enigma of Enset Starch Fermentation in Ethiopia: An Anthropological Study." In *Cured, Fermented and Smoked Foods: Proceedings of the Oxford Symposium on Food and Cookery 2010,* edited by Helen Saberi, 106–20. Totnes, UK: Prospect.

Gebeyehu, Kebede Adane, and Jon Abbink. 2022. "Land, Sugar and Pastoralism in Ethiopia: Comparing the Impact of the Omo-Kuraz Sugar Projects on Local Livelihoods and Food (In) Security in the Lower Omo Valley." *Pastoralism: Research, Policy and Practice* 12 (32). https://doi.org/10.1186/s13570-022-00242-8.

Glover, Jerry D., Cindy M. Cox, and John P. Reganold. 2007. "Future Farming: A Return to Roots?" *Scientific American* 297 (2): 82–89. https://doi.org/10.1038/scientificamerican0807-82.

Guinand, Yves, and Dechassa Lemessa. 2001. *Wild Food Plants in Southern Ethiopia: Reflections on the Role of "Famine Foods" at a Time of Drought.* United Nations Development Program (UNDP), Emergencies Unit for Ethiopia (UNDP-EUE), Rome.

Hadero, Haleluya. 2019. "Planting Millions More Trees Might Not Be the Way to Tackle Ethiopia's Environmental Problems." *Quartz Africa,* August 1. https://qz.com/africa/1679528/ethiopias-tree-planting-might-not-help-its-climate-change-battle/.

Harrison, Rodney. 2017. "Freezing Seeds and Making Futures: Endangerment, Hope, Security, and Time in Agrobiodiversity Conservation Practices." *Culture, Agriculture, Food and Environment* 39 (2): 80–89.

Hartigan, John. 2017. *Care of the Species: Races of Corn and the Science of Plant Biodiversity.* Minneapolis: University of Minnesota Press.

Hunduma, Teshome. 2020. "GMO Debate Is Democratic Test for Liberalizing Ethiopia." *Ethiopia Insight*, June 3. https://www.ethiopia-insight.com/2020/06/03/gmo -debate-is-democratic-test-for-liberalizing-ethiopia/.

Igoe, Jim, and Dan Brockington. 2016. "Neoliberal Conservation: A Brief Introduction." In *The Environment in Anthropology: A Reader in Ecology, Culture, and Sustainable Living*, eds. Nora Haenn, Richard R. Wilk, and Allison Harnish, 324–31. 2nd ed. New York: NYU Press.

Jacobsen, Kim, Guy Blomme, K. Tawle, Sadik Muzemil, and Zerihun Yemataw. 2018. "Dietary Diversity Associated with Different Enset [Ensete ventricosum (Welw.) Cheesman]-based Production Systems in Ethiopia." *Fruits* 73 (6): 356–64.

Jerneck, Anne, and Lennart Olsson. 2013. "More than Trees! Understanding the Agroforestry Adoption Gap in Subsistence Agriculture: Insights from Narrative Walks in Kenya." *Journal of Rural Studies* 32: 114–25. https://doi.org/10.1016/j .jrurstud.2013.04.004.

Kippie Kanshie, Tadesse. 2002. *Five Thousand Years of Sustainability? A Case Study on Gedeo Land Use*. Heelsum: Treemail.

Kirksey, S. Eben, Nicholas Shapiro, and Maria Brodine. 2013. "Hope in Blasted Landscapes." *Social Science Information* 52 (2): 228–56.

Kumar, B. M., and P. K. R. Nair, eds. 2006. *Tropical Homegardens: A Time-Tested Example of Sustainable Agroforestry*. Vol. 3 of *Advances in Agroforestry*. Dordrecht: Springer Netherlands.

Lemenih, Mulugeta, and Habtemariam Kassa. 2014. "Re-Greening Ethiopia: History, Challenges and Lessons." *Forests* 5 (8): 1896–1909. https://doi.org/10.3390/f50 81896.

Lewis, Simon L., Charlotte E. Wheeler, Edward T. A. Mitchard, and Alexander Koch. 2019. "Restoring Natural Forests Is the Best Way to Remove Atmospheric Carbon." Comment, *Nature*, April 2. https://www.nature.com/articles/d41586-019 -01026-8.

Lounela, Anu, Eeva Berglund, and Timo Kallinen, eds. 2019. *Dwelling in Political Landscapes: Contemporary Anthropological Perspectives*. Helsinki: Suomalaisen Kirjallisuuden Seura.

Lund, Ragnhild, and Axel Baudouin. 2023. "Neocolonial Agenda: Agrarian Transformations in Ethiopia and Sri Lanka." *Norsk Geografisk Tidsskrift / Norwegian Journal of Geography* 77 (2): 114–29.

MacEntee, Katie, Jennifer Thompson, Sirawdink Fikreyesus, and Kemeru Jihad. 2013. "'Enset Is a Good Thing': Gender and Enset in Jimma Zone, Ethiopia." *Ethiopian Journal of Applied Sciences and Technology*, special issue no. 1, 103–9.

McCann, James C. 2007. *Maize and Grace: Africa's Encounter with a New World Crop, 1500–2000*. Cambridge, Mass.: Harvard University Press.

Mellisse, Beyene Teklu, Gerrie W. J. van de Ven, Ken E. Giller, and Katrien Descheemaeker. 2018. "Home Garden System Dynamics in Southern Ethiopia." *Agroforestry Systems* 92 (6): 1579–95. https://doi.org/10.1007/s10457-017-0106-5.

Nair, P. K. R. 2001. "Do Tropical Homegardens Elude Science, or Is It the Other Way Around?" *Agroforestry Systems* 53:239–45. https://doi.org/10.1023/A:10133888 06993.

Nair, P. K. Ramachandran. 2017. "Managed Multi-Strata Tree + Crop Systems: An Agroecological Marvel." *Frontiers in Environmental Science* 5: art. 88. https://doi.org/10.3389/fenvs.2017.00088.

Negash, Almaz, and Anke Niehof. 2004. "The Significance of Enset Culture and Biodiversity for Rural Household Food and Livelihood Security in Southwestern Ethiopia." *Agriculture and Human Values* 21:61–71.

Orent, Amnon. 1979. "From the Hoe to the Plow: A Study in Ecological Adaptation." In *Proceedings of the Fifth International Conference on Ethiopian Studies, Session B*, edited by Robert Hess, 187–94. Chicago: University of Chicago Press.

Pearce, Fred. 2019. "Why Green Pledges Will Not Create the Natural Forests We Need." *Yale Environment 360*, April 16. https://e360.yale.edu/features/why-green-pledges-will-not-create-the-natural-forests-we-need.

Perfecto, Ivette, John Vandermeer, and Angus Wright. 2009. *Nature's Matrix: Linking Agriculture, Conservation and Food Sovereignty*. London: Earthscan.

Peveri, Valentina. 2020. *The Edible Gardens of Ethiopia: An Ethnographic Journey into Beauty and Hunger*. Tucson: University of Arizona Press.

Peveri, Valentina. 2021. "Flavoring the Nation: The Rhetoric of Nutrition Policies in Ethiopia." In *Rhetoric and Social Relations: Dialectics of Bonding and Contestation*, edited by Jon Abbink and Shauna LaTosky, 253–82. New York: Berghahn Books.

Rahmato, Dessalegn. 1995. "Resilience and Vulnerability: Enset Agriculture in Southern Ethiopia." *Journal of Ethiopian Studies* 28 (1): 23–51.

Scoones, Ian, and Andy Stirling. 2020. *The Politics of Uncertainty: Challenges of Transformation*. London: Routledge.

Scott, James C. 1998. *Seeing Like a State: How Certain Schemes to Improve the Human Condition Have Failed*. New Haven, Conn.: Yale University Press.

Shack, William A. 1971. "Hunger, Anxiety, and Ritual: Deprivation and Spirit Possession Among the Gurage of Ethiopia." *Man*, n.s., 6 (1): 30–43.

Sheridan, Michael. 2016. "Boundary Plants, the Social Production of Space, and Vegetative Agency in Agrarian Societies." *Environment and Society* 7:29–49.

Smith, Joe, and Petr Jehlička. 2013. "Quiet Sustainability: Fertile Lessons from Europe's Productive Gardeners." *Journal of Rural Studies* 32:148–57.

Trapp, Micah M. 2016. "YOU-WILL-KILL-ME BEANS: Taste and the Politics of Necessity in Humanitarian Aid." *Cultural Anthropology* 31 (3): 412–37.

Tsegaye, A., and P. C. Struik. 2002. "Analysis of Enset (*Ensete Ventricosum*) Indigenous Production Methods and Farm-Based Biodiversity in Major Enset-Growing Regions in Southern Ethiopia." *Experimental Agriculture* 38:291–315.

van der Veen, Marijke. 2014. "The Materiality of Plants: Plant–People Entanglements." *World Archaeology* 46 (5): 799–812. https://doi.org/10.1080/00438243.2014.953710.

CHAPTER EIGHT

A Tale of Two Rices

JUSTIN SIMPSON

The 2019 Intergovernmental Panel on Climate Change's special report found that agriculture, forestry, and other types of land use are responsible for 23 percent of human greenhouse gas emissions. Climate change is expected to exacerbate environmental problems related to agriculture, such as pollution, the encroachment of farms into forests, desertification, and the exhaustion of freshwater sources. With so much at stake, it is worthwhile to reconsider the ethical discussion surrounding food and agriculture. Two varieties—golden rice and Carolina gold rice—provide a window into food and agricultural studies. Golden rice (a genetically modified variety of *Oryza sativa* or Asian rice) has become a poster child of genetically modified crops, exemplifying the humanitarian promise of technological solutions to food insecurity and agricultural problems. Dating back to the eighteenth century, the landrace Carolina gold (a variety of *Oryza glaberrima* or African rice) has witnessed a renaissance on farms and plates in the American South. Golden rice and Carolina gold offer competing visions for the future of food—one based on productivism and disembeddedness, the other on idiosyncrasy and embeddedness. Jettisoning a dualistic and atomistic framework, this chapter draws on performative new materialism to shift the focus from individual varieties' biological characteristics to the intra-actions of their human-nonhuman networks. Rather than the pure, natural biodiversity metric, I develop biocultural diversity as an ethical criterion to compare these rice networks.

Golden Rice and Carolina Gold Rice

Golden rice is an incisive window into the debate surrounding the future of food because it embodies the top-down, disembedded, and production-driven approach to agriculture and food. Exemplifying the spirit of productivism behind green revolution rices, golden rice prioritizes output at the expense of other agricultural considerations (Stone and Glover 2016, 88). Although Stone and Glover (2016) note that golden rice has yet to meet standards of productivity due to its comparatively lower potential crop yield, the development of golden rice nevertheless prioritized maximizing the plant's nutritional output. By inserting two daffodil genes and one bacterium gene into the rice genome, European biologists with American funding created golden rice to be a biofortified crop with higher levels of beta-carotene, which is a precursor to vitamin A. This work sought to address the problem of vitamin A deficiency in food-insecure countries such as the Philippines. Severe vitamin A deficiency can cause blindness and even death in children as well as pregnancy complications (Sandler 2005). Golden rice thus represents a specialized technological solution to nutritional deficiency and food insecurity. Because of its humanitarian potential, golden rice has become a poster child of genetically modified crops.

In addition to productivism, golden rice embodies disembeddedness. Stone and Glover (2016) define embeddedness as "the extent to which local agro-ecological context is valorized or nullified in the crop's construction" (Stone and Glover 2016, 88). Yet, genetic engineering attempts to hardwire desired properties into plants at the genetic level. Such a process is premised on a framework that reduces a plant's complex biological processes as a living organism to genotypes, germplasm, gene constructs, and individual genes. Insofar as the plant is reduced to genes, and the desired properties are genetically built into the crop itself, this assumes a division between the individual organism and environment. This assumption conceals how the plant as a whole and its desirable properties are dependent on environmental interactions and management techniques (Charles 2001; McAfee 2003). Accordingly, although primarily intended for the Philippines, the developers did not originally breed golden rice with local varieties. Rather, they pursued the cosmopolitan breeding strategy of green revolution rices that resulted in golden rice becoming an amalgamation of rice DNA taken from various locations and inserted into green revolution rice (Harwood 2015; Richards 1997, both cited in Stone and Glover 2016). Starting in 2001, golden rice arrived at the International Rice Research Institute (IRRI) as the *javanica* subspecies, a U.S. commercial variety. It was then cross-bred with *indica* varieties commonly grown in the Philippines (Dubock 2014). In the early

2000s, researchers crossed the trait responsible for beta-carotene into Rc82 and the green revolution standby IR-64.

Proponents of genetic engineering present the disembeddedness of golden rice as a virtue. Since the nutritional properties are supposedly genetically encoded, golden rice's cultivation success is not contingent on place-specific environmental growing conditions. Even if golden rice's reported success and yields during testing at IRRI depended on ideal rice plots with placeless chemical fertilizers and modern technology, the intended growing site could theoretically reproduce these conditions—assuming that local farmers can afford to do so. Furthermore, proponents predict that its success is not dependent on cultural factors. Since the "technology is in the seed, no manufacturing, packaging, distribution, or change of cultural practices is required for populations to improve their nutritional status" (Dubock 2014, 214). Hence, golden rice could be a solution to vitamin A deficiency that can be grown in other locations or exported worldwide.

While scientists developed golden rice, there was a renaissance of the antebellum landrace Carolina gold rice. As a landrace, Carolina gold was bred using preindustrial domestication methods and has since been maintained agriculturally, rather than by scientific breeding in conjunction with modern farms (Roberts 2011). Carolina gold took root on plantations in the American South during the eighteenth century (Carney 2001; Tibbetts 2006). Commercial interest in Carolina gold grew as a cash crop to export back to Europe. The early success of Carolina gold, and the cultivation of rice in general as a southern staple, depended not only on the labor of enslaved people but also on their expertise. They processed the unmilled seed on slave ships, planted it in their personal plots as a dietary staple, and made it a viable crop in the colonies using West African agricultural techniques and knowledge (Carney 2001).

Consequently, after the Civil War, the prominence of Carolina gold declined and then disappeared after the devastating hurricane of 1911. Carolina gold returned on the southern scene in the 1980s with duck-hunting enthusiast Richard Schulze, who planted the rice as food after hearing about the exceptional taste of ducks fed on Carolina gold (Schulze 2005). By 2000, Carolina gold became viable for growing as an exotic crop in a global rice market saturated with inexpensive rice from overseas. Carolina gold regained its public prestige with the help of southern farmers like Glenn Roberts and famous regional chefs like South Carolina's Sean Brock.

In contrast to golden rice's productivism and disembeddedness stands the idiosyncrasies and embeddedness of landraces like Carolina gold. Landraces display individual variation at the genetic level. This variation is "celebrated and encouraged" by Roberts for its survival advantages, such as the ability "to maintain crop vigor, success in low-resource, high-stress environments . . . drought resistance, disease and pest tolerance . . . and the innate ability to adapt to climate and other change" (2011). Landrace advocates also prize the idiosyncrasies for their agronomic traits. Carolina gold's deep roots reflect its height. This height allows growers to use Carolina gold for thatch, bedding, fodder, and silage. Additionally, some landraces have allelopathic properties, such as suppressing weeds without tillage, which makes them more energy efficient, while others can suppress pests and have medicinal properties. Moreover, proponents appreciate Carolina gold for culinary and nutritional reasons. According to Roberts, it is the individual variation of landraces that "contributes to the appealing and dramatic flavor profile and high-quality nutrition" (Roberts 2011). Carolina gold's nutritional content is related to its deep roots. As Roberts explains, these roots allow it to "better uptake water and nutrients including important micronutrients for human and animal nutrition" (Roberts 2011).

Carolina gold's variety and embeddedness is also witnessed in the fields of Anson Mills' farms. Roberts practices polyculture farming, planting cover crops such as runner peas to block sunlight and minimize weeds and tilling. Roberts has also adopted sun cycle rotation, a type of crop sequencing. By planting Carolina gold after field peas, Roberts noticed an improvement in the rice flavor and soil tilth. Finally, variety is found in Roberts's harvesting process. Carolina gold is not only a fragile crop that easily falls over under pressure; it also has a variable maturation rate. Accordingly, it resists uniform and large-scale harvesting methods using industrial machines. Roberts therefore often turns to hand harvesting fields multiple times. Harvesting techniques can vary from manual artisan handwork to machine assistance.

Beyond the rich history of landraces, the preservation of local culinary traditions drives Roberts's work. Carolina gold has gained notoriety through southern restaurants like Sean Brock's Husk. Brock uses the crop to recreate flavorful traditional southern recipes, like the signature dish of South Carolina—hoppin' John—a stew consisting of rice, black-eyed peas, and pork. Regional restaurants like Husk pride themselves on using traditional recipes and ingredients. These recipes and ingredients are combined with cooking methods such as the whole-animal approach to cooking, which utilizes the entire animal instead of only the prized cuts.

Like the rice itself, these southern recipes and culinary methods are indebted to enslaved people and their descendants. Segregation was a ubiquitous feature of the South, except when it came to food (Tibbetts 2006). Creole food is the product of the intermingling of European and African food cultures. While enslaved people prepared European dishes for plantation owners using what produce was available, they did so by drawing on African cooking techniques, seasoning, and flavor profiles. This resulted in the creation of dishes such as hoppin' John. Additionally, soul food and the whole-animal approach to cooking were creative ways to supplement their diet with dishes consisting of wild fish, game, and animal parts discarded by whites, such as the feet, ears, heads, and entrails (Tibbetts 2006).

Although they are both rice, golden rice and Carolina gold present different "visions of what rice should be, and how it should be produced" (Stone and Glover 2016, 88). Each has garnered controversial acclaim as the future of agriculture and food. Against the backdrop of food crises, socioeconomic and cultural issues, and environmental problems, it is necessary to have an ethical criterion based on an ontological framework that can attend to how each rice is related to and affects these issues. For these reasons, I turn to performative new materialism (Gamble, Hanan, and Nail 2019). To articulate and motivate this account, the next section introduces the ontology associated with modernity as a point of contrast. I then show how this dualistic and atomistic ontology undergirds golden rice, conceptions of biodiversity, and the ethical framing of food and agricultural issues. The section concludes by identifying the problems with modernity, motivating the turn to performative new materialism.

Modernity

According to Bruno Latour (1993), modernity assumes that everything in the world can be sorted into the pregiven and fixed categories of culture/subject/human and nature/object/nonhuman. These categories represent a dualistic dichotomy and are mutually exclusive in that they do not share common properties. Accordingly, modernity regards nature-culture mixtures as primitive confusions and inappropriate projections. Works of purification seek to cut the world at these natural and cultural joints by severing inappropriate natural or social elements to sort entities into their proper original category. Making a similar point, Alfred Whithead contends that such projects correspond to a "bifurcation of nature into two systems of reality" that separate nature as it appears in human experience from the physical entities and processes that cause these experiences (Whitehead 1964, 30). Experiential secondary qualities, such as the "greenness of the trees, the song of the birds, the warmth of the sun, the hardness of chairs, and the feel of velvet," are discredited since they supposedly do not reflect the objective state of nature (Whitehead 1964, 31).

As a result of human/nonhuman dualism, nature is reduced to primary, quantifiable properties such as location, mass, volume, and velocity. Nonhumans consequently become passive inert matter that move according to deterministic physical laws. In Latour's words, nonhumans are regarded as mere "intermediaries" that transport external forces and information without transformation (Latour 2007). That is, individual nonhumans do not change anything or make a difference. Since their input is supposedly always identical with their output, they are interchangeable and predictable. Like with projectile motion, a person can predict the system's final state by knowing the initial state and applying the kinematic equations. Thus, as Arne Naess puts it, "The sphere of real facts is narrowed down to that of mechanically interpreted mathematical physics" (2001, 66).

Similar to works of purification breaking down mixed phenomena into their natural and cultural components, modernity breaks down relationships into atomistic units (Barad 2008). Reality is an aggregate of preexisting, discrete, and independent entities with intrinsic essences. Each entity occupies a definite location within a world, which is conceived as a fixed, external spacetime container. Within this contained-container model, there is always already distance between entirely separate entities. As independent variables, humans and nonhumans, a thing and its environment, can be neatly separated from each other and analyzed in isolation without loss (Tuana 2008). While these entities can be destroyed, combine, and interact in different ways, such as causing changes in an entity's mass, location, or momentum, these interactions cannot substantially change an entity's being, spacetime, or causality. The basic building blocks of the world, and their corresponding possibilities, thereby remain more or less given and fixed.

The legacy of modernity is observed in golden rice's abstract and disembedded design model. Atomism justifies the cosmopolitan crop design model of cutting and combining desirable genetic "atoms" from different organisms. Atomism also engenders the view that rice can be neatly severed from biological processes, the environment, and agricultural practices. Golden rice ultimately represents a universal, atomistic solution. Since its nutritional properties are supposedly inherent, being genetically encoded, it becomes an encapsulated vitamin supplement. It can therefore be designed in one place and grown on farms in another place without significant problems or disruptions to the farmers, agricultural animals, or local ecosystems.

Accounts of biodiversity also reflect modernity's dualism between humans and nonhumans. Sarah Whatmore notes how the Convention of Biological Diversity "casts [biodiversity] in wholly biological terms, the outcome of an evolutionary process divested of human presence," which "conjures a world until recently unmarked by the (invariably negative) 'impacts' of human society, only countenancing hybridity as a technological accomplishment associated with the advent of genetic resources" (2002, 92). Such an account presents biodiversity as a pure, natural metric. While species are interdependent, this is still an atomistic conception insofar as they have fixed, inherent, and independent essences in the form of DNA and biological life cycles (Ingold 2011). Biodiversity becomes a measure of the number of discrete, different species that are defined purely in objective and quantifiable terms (Shiva 2012). Furthermore, as Whatmore notes, this positions biodiversity as not only independent from humans, but also in an antagonistic relationship with humans.

Additionally, modernity often frames ethical debates around genetically modified crops and agriculture. These debates center on whether crops are unnatural, are nutritionally inferior, pose risks to human health by introducing new toxins or allergens, or threaten biodiversity by outcompeting wild counterparts (Moghissi, Pei, and Liu 2015; Rauch 2018; Sandler 2005). Modernity's logic of purity (Lugones 1994; Shotwell 2016) is witnessed in the distinction between "natural" and "unnatural," as well as in the focus on the effects of crops on preexisting people and the land, which assume that these categories are substantially discrete, independent, and self-contained. This framing neglects how crops and agriculture can transform the being of humans, nonhumans, society, and the land. More generally, the logic of modernity is witnessed in food ethics when food crises are reduced to production and distribution problems, which result from a lack of availability, accessibility, utilization, and stability of nutritious food (FAO 2006). This logic reduces the rich and diverse forms of human and nonhuman life to quantitative terms such that solutions to food insecurity become merely a distribution problem of raw material resources that provide nutritional content.

While there are many problems with modernity's atomism and dualism, given the focus of this chapter, it is worthwhile to emphasize three in particular. The first involves practical difficulties. Research compiled by Stone and Glover (2016) reveals the challenges a messy, complicated amodern world presents to the viability of golden rice. For example, the highly cited work of Tang et al. (2012), which found that golden rice is a good source of vitamin A, was based on trials involving children with well-balanced diets. This research, however, did not consider how children suffering from malnutrition in food-insecure areas would have diets low in fats, which are required to absorb vitamin A (Dawe, Robertson, and Unnevehr 2002; Haskell 2012; Nestle 2001). Furthermore, this demographic of children would also reduce golden rice's actual efficacy because they are more prone to intestinal infections and parasites. Such conditions hinder the absorption of beta-carotene and decrease its conversion rate into vitamin A (Nestle 2001). Additionally, Tang's study neglects the effects of storage and cooking on beta-carotene levels. Yet, oxygen, light, and heat are known to break down carotenoids (Hansen 2013; Pénicaud et al. 2011). Finally, the pathway opened for beta-carotene might negatively affect the flow of other nutrients (Stone and Glover 2016). This relationship illustrates how golden rice's nutritional properties are not intrinsic but depend on various factors, such as diet, the human body, storage, and cooking.

Second, modernity encourages an indifferent and violent relationship to nature (Plumwood 1993; Warren 1990). Understanding nature as a collection of passive, inert, and meaningless matter strips nonhumans of their sovereignty. In other words, with this understanding, nonhumans lack their own organizing principles, interests, subjectivities, agencies, and forms of intelligence. Modernity's nature consequently beckons humans to inscribe it with meaning by manipulating, developing, and exploiting nature to serve (some) human purposes. There is an ethical imperative to use nature. In rendering nonhumans as merely raw material resources with instrumental value, modernity fails to recognize the sovereignty of nonhumans, which can impose ethical restrictions on human actions.

Third, equating nature with only objective, quantifiable properties undercuts environmental ethics in two additional ways. On the one hand, modernity divorces humans from nature and delegitimates personal, qualitative relationships with nature. In Arne Naess's words, "Every appeal to save parts of nature based on reference to sense-qualities of any kind become meaningless. Every passionate appeal that involves deep feelings, empathy, and even identification with natural phenomenon must then be ruled as irrelevant" (2001, 65–66). Environmental calls to action consequently lose their affective force and are reduced to appealing to the cold hard facts and numbers. On the other hand, this physical and quantitative representation of nonhumans can elicit a cold indifference to nonhuman suffering and oppression as well as environmental destruction. "The framework of reductive mechanism," Val Plumwood contends, "permits the emotional distance which enables power and control, killing and warfare, to seem acceptable, just as it did in the case of the animals Descartes' followers used for experimentation" (1993, 119). These practical and environmental concerns motivate moving beyond the modern framework.

Performative New Materialism

The ontology of performative new materialism offers one antidote to the problematic metaphysics of modernity. Rather than starting with a closed set of discrete, self-contained humans and nonhumans that are given from the start and subsequently *interact*, Karen Barad's performative new materialism presents the world as a contingent, open-ended but nonarbitrary process of becoming. Humans and nonhumans are the product of *intra-actions*, which consist of actions and practices. Intra-actions are "generative material articulations or reconfigurings of what is and what is possible" (Barad 2007, 389). Particular intra-actions materialize and coconstitute humans and nonhumans in particular ways. Each is a part of the world, being affected by the world and affecting it.

According to Barad, even scientific practices and theories do not merely *re*present and *re*produce nature as it is, independent of humans. In other words, "Nature is not a pure essence that exists 'out there'" waiting to be discovered (Barad 2007, 382). As a part of the world, scientific experiments affect the world. Like all intra-actions, scientific experiments are performative in that "what is 'disclosed' is the effect of our intra-active participation with/in the world's differential becoming" (Barad 2007, 361). Following Niels Bohr's complementarity principle in quantum physics, Barad contends that particles and light do not have a preexisting determinate state that is simply unknown before the experiment. Prior to the experiment there is not only an epistemic indeterminacy, but also an ontological indeterminacy. Light or a particle only becomes locally determinate in relation to the particular material-discursive configuration of the scientific measuring apparatus. With a movable apparatus, light is a wave; with a fixed and rigid apparatus, light is a particle. Scientists are therefore not completely detached (and innocent) observers. The very method and material-discursive apparatuses that make possible the observation can substantially affect what is observed. To be clear, though, Barad is not advocating a new form of idealism or social constructivism. Intra-actions are not determined solely by humans since "phenomena do not require cognizing minds for their existence" (Barad 2007, 361).

More specifically, intra-actions involve agential cuts that enact local ontological resolutions in the world. This cut is an ontological performance insofar as it constitutes the very being of entities, giving them particular bodily boundaries, meanings, and agential abilities. An organism's bodily practices differentiates itself from the environment and differentiates its environment in particular ways that matter to it, while excluding others from mattering. For instance, a plant's activity distinguishes between sun, air, water, soil, rocks, and roots (Baluška et al., 2009; Gerber and Hiernaux 2022). That is, a particular intra-action enacts a cut in the world that separates the organism from the environment. The organism/environment division, or for that matter human/nonhuman division, is not given in advance but enacted by the intra-active practice.

While agential cuts enact differences and separation, they do not result in the absolute exteriority of atomism. Intra-actions cut things apart, but also cut things together. Intra-actions involve a process of differential articulation of the observed agency by the agency of observation. The resultant entities are consequently an "entangled relation of difference" (Barad 2007, 236). They remain connected even if they are not necessarily physically contiguous. For example, since the atom-as-particle would not be what it is without the corresponding fixed and rigid measuring apparatus, the "atom includes the apparatus that helps constitute" (2007, 472).

A similar point can be made about rice. As Stone and Glover maintain, rice is more than individual grains. Rice is always a "rice world" that "encompasses biological characteristics of seeds, agro-ecologies of rice landscapes, processing, and marketing, key institutions, visions of what rice should be,

and how it should be produced" (2016, 88). These more-than-human associations function together as a material-discursive apparatus, coconstituting
what each rice is. These associations are not simply exterior additions resulting in dissectible mixtures. They are enfolded into the being of each rice
such that each rice is entangled with them. As Haraway explains, "A seed
contains inside its coat the history of practices," which can include collecting, breeding, planting, calendars, labor systems, pest-control procedures,
landholding, marketing, taxonomizing, patenting, biochemically analyzing,
advertising, eating, cultivating, harvesting, celebrating, and beliefs about
hunger and well-being (1997, 129). Furthermore, the agency associated with
its nutritional content is not a fixed and intrinsic property of golden rice
itself. It is a network phenomenon, which is to say in between, relational
properties. Whether golden rice is a good source of vitamin A depends on
the recipient's diet, intestinal infections, parasites, storage, and cooking.

Of course, intra-actions cut both ways. Barad seeks to recognize nonhuman agency and how nonhumans are active participants in intra-actions.
Such a recognition, however, is not a motion for "uncritical equality" that extends human agency to nonhumans (Plumwood 1993, 27). Nonhumans have
different kinds of agency. Furthermore, moving beyond human-centered accounts of agency, Barad contends that "agency is about changing possibilities
of change entailed in reconfiguring material-discursive apparatuses of bodily
production" (2007, 235). This agency does not merely involve actualizing
one possibility among a given set but generating new sets of possibilities.
Making a similar albeit different point, Latour explains how nonhuman
agency refers to how nonhumans are "able to propose their own theories of
action to explain how agencies' effects are carried over [having] their own
meta-theory about how agency acts. . . . They will not only enter into a con-
troversy over which agency is taking over but also on the ways in which it is
making its influence felt" (2007, 57).

Consequently, insofar as nonhumans are active participants in the world,
nonhumans can reconfigure the being of humans—our bodily boundaries,
meaning, and agential abilities. Like objects, human subjects are "permeated
through and through with their entangled kin; the other is not just in one's
skin, but in one's bones, in one's belly, in one's heart, in one's nucleus, in
one's past and future" (2007, 393). The agency of nonhumans extends beyond constituting individual humans. As Barad notes, "The entanglements
we are part of reconfigure our beings, our psyches, our imaginations, our
institutions, our societies" (2007, 383). Nonhumans are therefore not passive

external objects "out there," but also in here—in us. As Emanuele Coccia puts the point, "Each one of us is a little zoo that always carries many more species than the one to which we imagine we belong" (2021, 126).

Nonhuman agency further evinces the sovereignty of nonhumans. For instance, the sovereignty of rice pertains to its own material, bodily structure, and biological processes as a living organism. From the way it grows, its favorable environmental conditions, when and how to harvest, how to store, how to cook, what types of dishes to cook, and what to serve it with, rice is a material-discursive apparatus that has its own material terms and dynamics, which translate, coconstitute, and organize the rest of the world around it. Through iterative intra-actions, these dynamics become differentially enfolded into the being of humans. They shape and translate human bodies, experiences, thinking, beliefs, values, intentions, practices—both on and off the field. Each world of rice consequently opens up particular human possibilities and futures while simultaneously foreclosing others.

This sovereignty is not limited to groups of nonhumans. Even nonhumans of the same type or species can have different materialities, agential abilities, intelligences, and ways of life. Like humans, each is a singular entity insofar as it has a specific history of intra-actions with others. This history is "written into their materialization, their bodily materiality holds the memories of the traces of its enfoldings" (Barad 2007, 383). Consequently, nonhumans are not necessarily intermediaries, but can also be mediators. As Latour defines it, a mediator is a "unique event, totally irreducible to any other" (2007, 153). It can make a difference in unexpected ways by translating forces, transforming information flows, and transforming other entities. Since each entity has its particular differential materiality, spacetime, agential abilities, and causal theory that translate action, "their input is never a good predictor of their output" (Latour 2007, 39). Acknowledging the sovereignty of each nonhuman, both in terms of their group affiliation and as individuals, is thus a matter of recognizing how each nonhuman is a mediator that makes a difference in the world according to their own terms.

Biocultural Diversity

With performative new materialism, it is necessary to reconceive and expand the modern concept of biodiversity with the notion of biocultural diversity (Baker 2013; Kimmerer 2013). This concept captures the particular entangled relations of difference between humans and nonhumans that result from contingent, iterative, and open-ended intra-actions. Biocultural diversity captures the qualitative diversity of nonhumans. More than the number of distinct biological species, biocultural diversity attends to the different materialities, agential abilities, interests, intelligences, causal theories, spacetimes, and worlds of different species. Furthermore, biocultural diversity recognizes the qualitative diversity within the same category. An individual organism is not necessarily identical with that of another organism of the same species since it can change through its specific history of intra-actions. Thus, biocultural diversity registers the singularities of nonhumans and how they are nonsubstitutable mediators that affect the world's becoming in unique ways. Biocultural diversity also allows for nonhuman diversity to increase through the iteration of intra-actions and the world's open-ended process of becoming.

On the other side of the entanglement, biocultural diversity takes into account how humans are coconstituted through intra-actions with more-than-humans. Intra-actions fold nonhumans into the being of humans in particular ways. Intra-actions involving different nonhumans will produce humans with different bodies, minds, cultural values, and practices. Although humans are one species, biocultural diversity recognizes the qualitative differences within humans' being, which in turn coconstitute nonhumans in different ways.

Biocultural diversity also seeks to attend to the sensorial, affective, and emotional dimensions of nonhumans. As Whitehead argued, if we acknowledge that humans are part of nature, we cannot discredit the apparent nature of human experience. In his words, "We cannot pick and choose. For us the red glow of the sunset should be as much a part of nature as are the molecules and electric waves by which men of science would explain the phenomenon" (1964, 29). According to Naess, secondary properties' relational nature is not sufficient to exclude them from nature since even primary properties are relational claims. The objective measurement of "height" depends on another measuring device with a specific unit of measure. Rather than being subjective, these characteristics are "like smell, *bound in an interdependent relationship* to our conception of the world" (2001, 48). Additionally, Naess argues that feelings and emotions are a part of nature since they are integral parts of gestalts that frame experience. "Reality as spontaneously experienced," Naess writes, "binds the emotional and the rational into indivisible wholes, the gestalts" (2001, 63). Experience accordingly consists of wholes like a tall, dark, somber tree. The tree's height, darkness, and somberness are all a part of reality.

In Barad's terms, sensibilities, feelings, and emotions are aspects of the measuring apparatus of an individual human body. Consequently, when humans intra-act with nonhumans, we enact cuts that coconstitute nonhumans with sensorial, affective, and emotional dimensions. Rice therefore is not just rice simpliciter. It is not necessarily simply a seed or staple crop. Its being can also include taste, texture, the meaning of a filling meal, nostalgia, the feeling of home, and hope for the future. Of course, rice is not simply defined by human affects. It is also defined by how it is affected by and affects other nonhumans.

In comparison with biocultural diversity, modern biodiversity is not only impoverished but impoverishing. Insofar as biocultural diversity attends to the different forms of life—different ways that matter matters—biodiversity enacts a loss in biocultural diversity, even without the extinction of a species. Becoming worlds are reduced to one given world with a fixed set of possibilities. Furthermore, the diversity of nature is further reduced as a result of how this "nature" historically corresponds to the agential cut of a white, Western, patriarchal, ableist, capitalist, colonial culture, predominately practiced by heterosexual males with particular technology. As Heather Davis and Zoe Todd (2017) note, colonialism corresponds to an imperial, universalizing logic that severs relations. When nonhumans are understood in an atomistic way and severed from their environment, one further undermines the recognition of nonhuman agency and intelligence since one cannot appreciate how nonhumans affect and are affected by their environment. Moreover, considering all matter—from microorganisms to mammals—as inert, passive, and mechanical further reduces biodiversity because it neglects the possibility of not only nonhuman agency but different nonhuman agencies, spacetimes, and causal theories. The little diversity and agency that remains is further stripped away by the division of observer and observed, which leads to disregarding nature's sensorial, affective, and emotive dimensions.

A Performative New Materialist Environmental Ethic

Performative new materialism ultimately thickens and troubles environmental ethics. There is more to consider ethically than simply promoting, caring, or respecting others' interests, freedom, and needs. It is necessary to also consider intra-actions, since what well-being, respect, and care entail depend on the being of humans and nonhumans, which are themselves the production of previous intra-actions (Johns-Putra 2013). Moreover, these intra-actions are not innocent. They "make the world intelligible in specific ways but also foreclose other patterns of mattering" (Barad 2007, 394). Since intra-actions are part of the world, intra-actions are structured by power imbalances along the lines of gender, race, class, ability, and nation-states. These power relations make certain things materialize and become visible in particular ways, foregrounding a specific set of interests, backgrounding others, and excluding others from mattering at all. Although intra-active practices are always exclusionary, not all practices are exclusionary to the same extent or in the same way. Hence, "What is needed is . . . an ethics of worlding" (Barad 2007, 392).

An environmental ethic based on performative new materialism seeks to protect and promote biocultural diversity. Human life and flourishing are dependent on nonhumans for material sustenance. As Chris Cuomo puts it, "Humans cannot flourish without other humans, ecosystems, and species," (1998, 74). Humans therefore depend on healthy ecosystems, which "usually include a high number of different species and forms of life" (Cuomo 1998, 134). Nonhumans do not merely provide the necessary inputs—food, air, water, shelter—for humans to survive. Our being, identity, and agency are coconstituted by nonhumans through studying nonhumans, making and using technology, and engaging with rocks, rivers, plants, animals, and ecosystems. These more-than-human associations subjectivize humans, helping make each person an individual by coconstituting their particular needs and personal desires, bodily experiences of pleasure and pain, well-being, what they care about, hope for, and imagine. Furthermore, more-than-humans are integral to human agency. Human agency is no longer an antagonistic zero-sum game between one person and other humans and nonhumans such that "whatever [comes] from the outside [is] deducted from the total sum of action allotted to the agents 'inside'" (Latour 2007, 215). These connections with more-than-human others make humans more articulate and competent, having the sensitivity and agential abilities needed to achieve their goals. Human flourishing is thus intimately bound up with biocultural diversity as "living-with [is] the only possible way to live-well" (Haraway 2016, 162). Biocultural diversity provides more qualitatively diverse entities for humans to form connections with, to become with, which can ultimately lead to richer and more robust individual lives and communities with different bodies, sensitivities, agential abilities, and futures. Biocultural diversity thus reflects how human flourishing is always an act of multispecies flourishing.

Yet, biocultural diversity is not necessarily intrinsically good (Gunnarsson 2013). It is not a license to litter and pollute, nor does it justify introducing any destructive, harmful, and oppressive technology and hybrid organisms for the sake of novelty. Despite these actions increasing diversity, the increase is only temporary and ultimately results in a loss of diversity in the long term. Intra-actions, interactions, and biocultural diversity should therefore not undermine the primary conditions necessary to continue living and becoming in different ways. Human becoming should perform diversity in a sustainable way—in a way that does not diminish qualitative diversity and the possibilities for multispecies becomings over time (Braidotti 2006). That is, biocultural diversity becomes an ethical value when it sows the seeds of possibility and empowers by increasing everyone's ability to affect and be affected.

For biocultural diversity to promote mutual, multispecies flourishing, human and nonhuman differences must be able to actually make a difference. There cannot be hierarchical networks with gross inequities of power between humans and nonhumans, and more specifically between certain groups of humans and nonhumans. Although the differences might still technically be present, they do not have to be represented and seriously considered within hierarchical networks. They can be disregarded, backgrounded, oppressed, subjugated, and dominated as mere intermediaries in a system constructed to satisfy (some) human interests. Thus, to enrich and empower humans and nonhumans in different ways, biocultural diversity requires more-than-human egalitarian networks that are structurally open.

While a human-nonhuman network rich in biocultural diversity will help to sustain and promote multispecies flourishing and open new, different ways of becoming, sustaining biocultural diversity depends on individual actors performatively enacting biocultural diversity. Performing biocultural diversity depends on attending to the singularities of sovereign more-than-human others, regarding them as mediators that can make unique differences in events. Accordingly, understanding others cannot simply rely on abstract and detached analogical reasoning or transcendent metaphysics.

When these methods dictate our mode of engagement and understanding, they stifle, domesticate, and erase the differences we are trying to attend to by forcing them into preconceived categories (Bergson 1998). Rather, understanding must be immanently acquired through bodily interactions and intra-actions that "follow the actors themselves" through their world-making activities on the ground (Latour 2007, 61). This will require sensitive apparatuses and practices that are open to different nonhumans and their corresponding materialities, agential abilities, and causal theories. Such apparatuses and practices provide the conditions for the possibility of the observed entity to demonstrate its difference in terms of how it mobilizes, translates, and transforms others and how it is mobilized, translated, and transformed by others. These differences include the ways in which it resists or refuses to be moved. Following Rosi Braidotti, to enact biocultural diversity it is therefore necessary to "experiment with resistance and intensity in order to find out what posthuman bodies can do" (2019, 99).

To perform biocultural diversity, humans must embody biocultural diversity. No longer vilified as the source of epistemic error, moral vice, or constraint, the human body becomes a productive site of understanding. Expressing a similar sentiment, Naess contends that a richer, more integral, and "joyful experiencing of nature is partially dependent upon conscious or unconscious development of a sensitivity for qualities" (2001, 51). The development of this sensitivity, however, should not be construed as a humanist project of perfection that develops and disciplines the human body in the image of a Eurocentric, masculine, heterosexual, ableist ideal. Given how different bodies materialize different worlds, homogenizing human bodies would enact a loss in biocultural diversity, universalizing one agential cut. Since nature is an open-ended, iterative process of becoming and the human body lacks a given, fixed essence, developing this qualitative sensitivity will require experimenting "with what we are actually capable of becoming" (Braidotti 2019, 92).

Such experiments are not solely individual nor intersubjective undertakings though. A person becomes a more bioculturally diverse body through heterogeneous practices with qualitatively diverse more-than-humans. These intra-actions rematerialize human bodies and subjectivities by translating them according to the sensitivities, agencies, and materialities of more-than-humans. These nonhumans provide supplementary souls that "literally lend you the shape of a psyche" (Latour 2007, 212). Such practices embody bio-cultural diversity in humans. The resulting hybrid human body has more refined and different sensibilities and agential abilities—different ways of seeing, feeling, thinking, valuing, and acting. These sensibilities and agen-tial capabilities subsequently make us more sensitive and open measuring apparatuses, which can follow the actors themselves and attend to the sin-gularities of nonhumans.

These practices and bodies also perform biocultural diversity through enacting other, new nonhumans. By deferring to the other, following the actor itself, and mapping the circulation of entities around it with sensitive and open hybrid minds and bodies, we are better positioned to cross into their metaphysics and understand their causal theories. In doing so, "we can detect many other entities whose displacements were barely visible before" (Latour 2007, 205). Based on this different metaphysics, we can perceive and understand others in other ways, which are overlooked by an isolated individual working within the confines of a pure human-centered metaphys-ics. Experimental practices thus perform biocultural diversity in that "lots of surprising *aliens* may pop up in between" the circulating entities (Latour 2007, 59).

This hybridization of the human body is simultaneously an ethical act of becoming. Instead of being an abstract, detached ethical value that humans must apply to a situation, embodying biocultural diversity makes humans more entangled, invested, and committed with nonhumans. Bioculturally diverse human bodies are more sensorially, affectively, and emotionally con-nected with more-than-human others. We are ontologically accountable to them. We feel the moral gravity of the singularity of their life and their sovereignty in our bones and our hearts. Being more response-able accom-panies this felt responsibility. We are not only more sensitive to the other's more-than-human differences—the specificity of their needs, desires, and well-being—but have the agential abilities to attentively respond to them in ways that more fully care for nonhumans and humans. Heterogeneous prac-tices involving open and sensitive hybrid human bodies that treat nonhu-mans as mediators are consequently more inclined to interactively protect, and intra-actively perform, biocultural diversity. Such practices represent an ethic of worlding that helps ensure mutual flourishing in a nonideal, noninnocent world. With this ethic in hand, we can now compare the two rice worlds.

The Moral Behind the Tale of Two Rices

A combination of different genes and the product of international collaboration, golden rice initially appears to be the epitome of bicultural diversity. Golden rice entangles nature and culture, and as Barad asks, "Isn't the undoing of the very idea of an inherent nature-culture boundary a useful tool, if not a prerequisite, for destabilizing sexism, racism, and homophobia and other social ills that are propped up by this dualism and its derivatives?" (2007, 368–69). Nonetheless, the matter is not so simple. It is necessary to critically examine what is undone and redone, and what is destabilized and what in turn is given stability. In terms of the ethical criterion of bicultural diversity, these questions become: Are humans and nonhumans treated as mediators? Is it a sustainable form of becoming? Which humans and nonhumans matter, in what ways, and which are excluded?

Although golden rice brings together genes from different sources, it assembles these genes with the sole intent of producing one desirable trait— increased beta-carotene. This human interest dominates the design behind golden rice, reducing a complex, biological organism to a raw material—an intermediary for vitamin A. The growing, harvesting, and processing also ensure golden rice remains a mere intermediary. Golden rice is thus a reflection of human interests and reduced to mattering as merely a human resource. In this sense, golden rice lacks *bio*cultural diversity.

Golden rice also lacks bio*cultural* diversity. Although golden rice centers human interests, there is an asymmetry within this centralization of agency. The cultural work of agriculture is disembedded from the human and nonhuman actors in the field: the local soil, crops, animals, environment, farmers, community, and culture. Agency becomes predominantly relocated to scientists in the laboratories of land-grant, knowledge-based institutions, IRRI's controlled fields, modern technology, synthetic inputs, and agricultural companies, which are laced with different political, academic, cultural, and economic interests. Moreover, insofar as the ideal rice plots with agrochemicals and modern machinery are the model to emulate, locally implementing this farming strategy actively seeks to minimize, or at least overlook, the agency associated with environmental interactions and management techniques.

By not attending to the historical entanglement of the biological and cultural variables at the nexus of food issues, golden rice does not protect

biocultural diversity by offering a real solution to these issues. This specialized technological magic bullet only addresses the problem's symptoms
rather than the causes of the food insecurity that precipitated the vitamin
A deficiency. For instance, if international socioeconomic interventions and
the development and productionist paradigm of monoculture agriculture
eroded biocultural diversity via deforestation and the depletion of soil fertility, and if the increasing growth of cash crops for export created nutritional
deficiency, then this is what needs to be addressed. Not only does golden rice
not address these underlying causes, but it could also perpetuate the problem by concealing and maintaining the responsible practices by providing
only a temporary, topical solution. Golden rice would thereby perform a net
loss in biocultural diversity over time by temporally neutralizing the loss of
forest, soil fertility, and diversity of crops on the farm.

Furthermore, the technical challenges encountered in embedding the
"golden" transgenic trait in local varieties jeopardize its potential humanitarian value and the work and resources invested in it. Although in development
since 2001, golden rice has not been commercially released to farmers in the
Philippines. Since this time, the Philippines implemented food supplements,
nutritional education, and a law requiring the fortification of common ingredients with vitamin A (Stone and Glover 2016). Vitamin A deficiency
decreased from 40.1 percent to 15.2 percent of the population between 2003
and 2008 (Food and Nutritional Research Institute n.d.). Thus, "[golden rice]
shrinks from a sure solution to a pervasive public health crisis of the developing world, to a possible dietary supplement in certain, unnamed, difficult-
to-reach spots in the Philippines" (Stone and Glover 2016, 95).

Additionally, golden rice might decrease biocultural diversity. The history
of green revolution rice raises concerns about how agricultural colonialism
can seed cultural and economic hegemony. According to Nick Cullather
(2010), getting high-yield varieties into the hands of farmers in developing
countries was not the sole concern. Green revolution rice varieties were a
means to inculcate the methods, technology, and synthetic inputs used and
sold by the West into Filipinos' thinking and agricultural practices. Stone
and Glover explain how the point "was not only to displace locally adapted
seeds, but to transform locally embedded cultivation and breeding practices
and even local food, cultures, and rural ways of life" (2016, 93). Similarly,
remarking on India's recent history with biotechnology, monoculture, and
the development paradigm, Vandana Shiva (2012) notes how they fail to

register, and ultimately erode, women's agricultural skills, knowledge, and social status.

While green revolution rice and golden rice have important differences, there is still reason to be concerned about the homogenizing effects of golden rice. Proponents claim that no "change of cultural practices is required" (Dubock 2014). But this raises the question: no change with respect to what? If the baseline is the green revolution, and golden rice yields depend on modern technology, agricultural practices, and chemical inputs, this still imperils certain human subjectivities, values, knowledges, and cultures that are sustained through different (nonmasculine, nonindustrial, nonproductionist) relationships with seeds, crops, technology, and the land. In addition, the reliance on imports would further wed populations to the global market. Historically, this global market has been structured to reinforce power imbalances such that the agency, subjectivities, and interests of some do not have to be seriously regarded in their own right. According to John Cobb Jr., "The intended and actual consequence of present economic policy is that no community or nation feeds, or houses, or clothes itself. All of this is to depend on trade. . . . They cannot live without importing the necessities of their livelihood, however unfavorable the terms of the trade may be" (2012, 686). These connections would diminish the sovereignty of golden rice farmers by making them less able to respond, and therefore more vulnerable, to environmental, economic, and sociopolitical changes.

On the other hand, from the perspective of Glenn Roberts and Sean Brock, the world of Carolina gold welcomes and celebrates biocultural diversity. Biocultural diversity is witnessed in the genetic lineage behind landraces and the genetic variations of individual plants. Beyond its nutritional content, Carolina gold rice affects humans on multiple sensuous levels. In the kitchen, the nutty, earthy aroma released while cooking had Keith Pandolfi (2016) craving the rice. At the table, Carolina gold's subtle green tea, nutty, and floral flavor led Sean Brock (2014) to claim that it was "the most flavorful rice I have ever tasted." The ducks that feed on Carolina gold rice might agree. Regardless, Carolina gold's agency left its mark on them, such as changing the way they taste. Carolina gold's uncommon starch character not only gives it a notably rich and chewy texture, but also allows it to emulate medium and short grain rice, lending it to a variety of dishes (Slow Food Foundation 2019). In addition to the agency associated with growing, cooking, and eating, Carolina gold can be used for thatch and fodder.

Moreover, biocultural diversity is also found in Roberts's fields of Carolina gold as a result of the practice of polyculture farming in conjunction with various technologies to grow, harvest, and mill Carolina gold. These fields themselves have agency in that their stunning golden hue aesthetically moves farmers and locals.

Moreover, this is not a world where humans unilaterally impose their interests. This world treats seeds, plants, the environment, technology, and humans more like mediators. Differences are not only recognized but encouraged to make a difference. The agricultural technology, methods, and practices that Roberts uses to grow, harvest, and mill are adapted to Carolina gold's idiosyncrasies. Additionally, the cooking techniques and the chefs themselves seek to serve as supporting actors that allow the rice to shine and speak for itself.

In general, there is an openness and sensitivity exhibited in a democratic negotiation process that takes into consideration the interests of crops, the ecology of the local environment, soil, the land, animals, farmers, consumers, community, and chefs. The result is a more sustainable world of multispecies flourishing. Carolina gold's resiliency to harsh environmental conditions and changes is good for farmers. Its greater nutrition and flavor are good for chefs and consumers. The local supply chains and distribution are good for local economies. Carolina gold also performs biocultural diversity by nourishing cultural identities, providing traditional ingredients to recreate regional dishes such as hoppin' John using traditional culinary techniques. Local supply chains, as well as whole-animal approaches to cooking, are also more environmentally sustainable. Furthermore, in addition to the rice-crop rotation system, Anson Mills' farms grow thirty sustainable acres of Carolina gold at Prospect Hill field on the Edisto River, one of the oldest tidal trunk and dike rice fields in the country. This rice farm thereby promotes biocultural diversity by improving water quality and helping to protect these environmentally threatened wetlands, which are a habitat for wildlife (Slow Food Foundation 2019).

As a result of this openness, the world of Carolina gold is not only "full of significant otherness," but also "full of kinds-in-the-making" (Haraway 2016, 12). Human actors like Roberts and Brock can treat Carolina gold as a mediator because they in various ways embody biocultural diversity. Carolina gold has created corresponding human sensibilities and agential capabilities. It literally makes these farmers and chefs by growing on and into their beings. They have become "non-natal kin" (Haraway 2016). They think and act in relation to it, immediately seeing through the lens of the more-than-human world of Carolina gold. It is in their bones, bodily dispositions, and hearts. They are sensitive and responsive to its unique characteristics, growing tendencies, and flavor profiles. This sensitivity opens up the space of agency for both humans and nonhumans. The same goes for the world of rice off the farm. While an homage to the region's culinary and agricultural heritage, southern food offers chefs and cooks the opportunity to creatively reinterpret recipes and experiment in becoming. Thus, Carolina gold both interactively protects and intra-actively performs biocultural diversity.

The world of Carolina gold is not without problems, though. In addition to a past marred by slavery, the celebrated world of Carolina gold risks decreasing biocultural diversity. Because of their notoriety, Roberts and Brock, both elite white males, disproportionally determine what tastes good and what are authentic, traditional southern ingredients, food, and cooking in the public imaginary (Jones 2017). Rather than being purely objective, biological, or historical, "taste and heritage are embedded in contemporary meaning-making and value-adding processes that serve particular economic and cultural ends" (2017, 219). For instance, the culinary judgments of Roberts and Brock are influenced by their personal histories, which subsequently produces selective views of the antebellum South and tasty southern cuisine. Thus, while there is a more symmetrical relationship between humans and nonhumans in Carolina gold riceworlds, this is only true of some humans, who perform a particular world. Moreover, the heritage and extraordinary flavor increase Carolina gold's price. Consequently, "consuming delicious Lowcountry heritage increasingly becomes a niche opportunity for the affluent and ostensibly enlightened" (Jones 2017, 232). Underscoring the injustice, Michael Twitty writes, "Our story has been used to raise the price point of many menus so much so that the descendants of the enslaved cannot afford to enjoy and appreciate the very edible heritage that was nourished by their Ancestor's skills, knowledge, and blood" (cited in Cadigan 2019).

If the world of Carolina gold is going to provide an egalitarian, bioculturally diverse vision of the future of rice, it is necessary to reckon with this crop's history and current inequities. To learn to "live and die well with each other in a thick present," Haraway writes, we must "make trouble, to stir up potent response to devastating events, as well as to settle troubled waters and rebuild quiet places" (2016, 1). Accordingly, when explaining Carolina gold, one cannot dismiss the history of slavery since this would be another way of "placing slaves and Africa in the background of rice history," and more generally, erasing the pivotal impact enslaved people had on America's economy, culture, and agriculture. (Carney 2001, 149). Staying with the trouble of this noninnocent varietal entails recognizing and centering the historical agency of enslaved people in Carolina gold's world. One cannot celebrate Carolina gold rice without recognizing the expertise and creativity of these people, which ultimately contributed greatly to the rice's biocultural diversity, agroecology, and southern culinary culture.

Insofar as critiques begin and end with white heterosexual males at the center, they risk repeating the mistakes of the past. Accordingly, staying with the trouble involves appreciating the thickness of the present in terms of Carolina gold's current biocultural diversity. Black chefs like Food Network star Kardea Brown, Benjamin "BJ" Dennis IV, and Mashama Bailey of The Grey restaurant in Savannah, Georgia, are also part of the worlding of Carolina gold and southern cuisine. Attending to these actors opens the space of agency, revealing a different southern world of rice with different more-than-human actors, ways of life, and relations to food. For instance, Dennis is a chef who "could not only acknowledge the Gullah Geechee origins of his dishes but also make those origins—and their present-day implications—his focus" (Cadigan 2019). Dennis's worlding goes from urban hubs like Charleston to the Sea Islands of South Carolina and Georgia and then back to Savannah and Brunswick, Georgia.

The Sea Islands are home to Gullah Geechee communities, originally formed by descendants of West African enslaved people who labored on antebellum rice plantations. Dennis's world of Carolina gold includes roadside restaurants, home kitchens, family farms like Jackie Frazier's Barefoot Farms, African American cultural institutions like Penn Center, the Taste of Gullah festival held at the historically Black Harrington School Cultural Center, and Bill Green's restaurant Gullah Grub, which doubles as a training center to teach local middle schoolers how to cook rice and fish, and grow vegetables (Cadigan 2019). Through his networking and seed sharing, "Dennis is able not only to anchor a people to their past, but bring back the kind of self-sufficiency that's always existed in this community. It's a small and simple act, but it's also a revolution" (Cadigan 2019). By centering such human actors, the world of Carolina gold becomes more diverse.

In summary, this chapter develops a performative account of biocultural diversity as an ethical metric to compare the worlds of Carolina gold and golden rice. Biocultural diversity sought to disrupt modern logics of purity associated with the nature/culture dualism, which has historically served as gold standard for judgment (Deleuze 1990). Instead, biocultural diversity directs attention to the human-nonhuman entanglements that result from specific, contingent intra-active practices of engagement. As an ethical metric, biocultural diversity promotes agential cuts and entanglements that produce sensibilities and response-abilities that sustain open-ended becoming. Such an ethical criterion was then applied to golden rice and Carolina gold to intimate which tale of rice is better and more wholesome. Attempting to avoid easy answers by using biocultural diversity as a standard to provide a definite assessment about whether either golden rice or Carolina gold is fool's gold, this criterion was used to stay with the trouble of each rice by revealing the problems that need to be addressed so that they do not compound historical injustices, ongoing violence and oppression, and the loss of biocultural diversity. Beyond these rice worlds, this ethical account helps envision and assess more wholesome and fuller futures of food. Such a future consists of worlds not premised on exclusion, exploitation, and oppression, but worlds replete with diversity, sustainable becoming, and multispecies flourishing.

References

Baker, Lauren. 2013. *Corn Meets Maize: Food Movements and Markets in Mexico.* Lanham, Md.: Rowman and Littlefield.

Baluška, František, Stefano Mancuso, Dieter Volkmann, and Peter W. Barlow. 2009. "The 'Root-Brain' Hypothesis of Charles and Francis Darwin." *Plant Signaling and Behavior* 4 (12): 1121–27.

Barad, Karen. 2007. *Meeting the Universe Halfway: Quantum Physics and the Entanglement of Matter and Meaning.* Durham, N.C.: Duke University Press.

Barad, Karen. 2008. "Posthumanist Performativity: Toward an Understanding of How Matter Comes to Matter." In *Material Feminisms,* edited by Stacy Alaimo and Susan Hekman, 120–56. Bloomington: Indiana University Press.

Bergson, Henri. 1998. *Creative Evolution.* Mineola, N.Y.: Dover.

Braidotti, Rosi. 2006. *Transpositions: On Nomadic Ethics.* Cambridge: Polity.

Braidotti, Rosi. 2019. *The Posthuman.* Cambridge: Polity.

Brock, Sean. 2014. *Heritage.* New York: Artisan.

Cadigan, Hilary. 2019. "People Say Gullah Geechee Culture Is Disappearing. BJ Dennis Says They're Wrong." *Bon Appétit,* August 14. https://www.bonappetit.com/story/bj-dennis-gullah-geechee.

Carney, Judith. 2001. *Black Rice: The African Origins of Rice Cultivation in the Americas*. Cambridge, Mass.: Harvard University Press.

Charles, Dan. 2001. *Lords of the Harvest: Biotech, Big Money, and the Future of Food*. Cambridge, Mass.: Perseus.

Cobb, John B., Jr. 2016. "Toward a Just and Sustainable Economic Order." In *Environmental Ethics: Readings in Theory and Application*, edited by Louis P. Pojman and Paul Pojman, 451–63. 6th ed. Boston: Cengage Learning.

Coccia, Emanuele. 2021. *Metamorphoses*. London: Polity.

Cullather, Nick. 2010. *The Hungry World: America's Cold War Battle Against Poverty in Asia*. Cambridge, Mass.: Harvard University Press.

Cuomo, Chris. 1998. *Feminism and Ecological Communities: An Ethic of Flourishing*. New York: Routledge.

Davis, Heather, and Zoe Todd. 2017. "On the Importance of a Date, or Decolonizing the Anthropocene." *ACME: An International Journal for Critical Geographies* 16:761–80.

Dawe, D., R. Robertson, and L. Unnevehr. 2002. "Golden Rice: What Role Could It Play in Alleviation of Vitamin A Deficiency?" *Food Policy* 27 (5–6): 541–60.

Deleuze, Gilles. 1990. "The Simulacrum and Ancient Philosophy." In *The Logic of Sense*, edited by Constantin Boundas, translated by Mark Lester and Charles Stivale, 253–65. New York: Columbia University Press.

Dubock, Adrian. 2014. "The Politics of Golden Rice." *GM Crops and Food* 5 (3): 210–22.

Food and Agriculture Organization (FAO) of the United Nations. 2006. "Food Security." Policy Brief 2. June.

Food and Nutritional Research Institute. n.d. *Seventh National Nutrition Survey, 2008–2009*. Department of Science and Technology (Philippines).

Gamble, Christopher, Joshua Hanan, and Thomas Nail. 2019. "What Is New Materialism?" *Angelaki* 24 (6): 111–34.

Gerber, Sophie, and Quentin Hiernaux. 2022. "Plant as Machines: History, Philosophy and Practical Consequences of an Idea." *Journal of Agricultural and Environmental Ethics* 35 (4): 1–24.

Gunnarsson, Lena. 2013. "The Naturalistic Turn in Feminist Theory: A Marxist-Realist Contribution." *Feminist Theory* 14 (1): 3–19.

Hansen, Michael. 2013. "Golden Rice Myths." GMWatch, August 28. http://gmwatch.org/index.php/news/archive/2013/15023-golden-ricemyths.

Haraway, Donna. 1997. *Modest_Witness@Second_Millennium.FemaleMan_Meets_OncoMouse: Feminism and Technoscience*. New York: Routledge.

Haraway, Donna. 2016. *Staying with the Trouble: Making Kin in the Chthulucene*. Durham, N.C.: Duke University Press.

Haskell, Marjorie J. 2012. "The Challenge to Reach Nutritional Adequacy for Vitamin A: β-carotene Bioavailability and Conversion—Evidence in Humans." *American Journal of Clinical Nutrition* 96 (5): 1193S–1203S.

Ingold, Tim. 2001. "From the Transmission of Representation to the Education of Attention." In *The Debated Mind: Evolutionary Psychology Versus Ethnography,* edited by H. Whitehouse, 113–53. Berg: Oxford.

Ingold, Tim. 2011. *Being Alive: Essays on Movement, Knowledge, and Description.* London: Routledge.

IPCC (Intergovernmental Panel on Climate Change). 2019. "Summary for Policymakers—Special Report on Climate Change and Land." IPCC. https://www.ipcc .ch/srccl/chapter/summary-for-policymakers/.

Johns-Putra, Adeline. 2013. "Environmental Care Ethics." *symplokē* 21 (1–2): 125.

Jones, Bradley M. 2017. "Producing Heritage: Politics, Patrimony, and Palatability in the Reinvention of Lowcountry Cuisine." *Food, Culture and Society* 20 (2): 217–36.

Kimmerer, Robin Wall. 2013. *Braiding Sweetgrass: Indigenous Wisdom, Scientific Knowledge, and the Teachings of Plants.* Minneapolis: Milkweed.

Latour, Bruno. 1993. *We Have Never Been Modern.* Cambridge, Mass.: Harvard University Press.

Latour, Bruno. 2007. *Reassembling the Social: An Introduction to Actor-Network-Theory.* New York: Oxford University Press.

Lugones, Maria. 1994. "Purity, Impurity, and Separation." *Signs: Journal of Women in Culture and Society* 19 (2): 458–79.

McAfee, Kathleen. 2003. "Neoliberalism on the Molecular Scale: Economic and Genetic Reductionism in Biotechnology Battles." *Geoforum* 34 (2): 203–19.

Moghissi, A. A., Shiqian Pei, and Yinzuo Liu. 2015. "Golden Rice: Scientific, Regulatory and Public Information Processes of a Genetically Modified Organism." *Critical Reviews in Biotechnology* 36 (3): 1–7.

Naess, Arne. 2001. *Ecology, Community and Lifestyle: Outline of an Ecosophy.* Cambridge: Cambridge University Press.

Nestle, Marion. 2001. "Genetically Engineered 'Golden' Rice Unlikely to Overcome Vitamin A Deficiency." *Journal of the American Dietetic Association* 101 (3): 289–90.

Pandolfi, Keith. 2016. "The Story of Carolina Gold, the Best Rice You've Never Tasted." *Serious Eats,* May 26. https://www.seriouseats.com/2016/05/carolina-gold-heir loom-rice-anson-mills.html.

Pénicaud, Caroline, Nawel Achir, Claudie Dhuique-Mayer, Manuel Dornier, and Philippe Bohuon. 2011. "Degradation of β-carotene During Fruit and Vegetable Processing or Storage: Reaction Mechanisms and Kinetic Aspects: A Review." *Fruits* 66 (6): 417–40.

Plumwood, Val. 1993. *Feminism and the Mastery of Nature.* London: Routledge.

Rauch, Jonathan. 2018. "Will Frankenfood Save the Planet?" *Atlantic,* October. https:// www.theatlantic.com/magazine/archive/2003/10/will-frankenfood-save-the -planet/302806/.

Roberts, Glenn. 2011. "Old School: Glenn Roberts Restores Carolina Grains." *Common Place: The Journal of Early American Life* 11 (3). https://commonplace.online /article/old-school/.

Sandler, Ronald. 2005 "A Virtue Ethics Perspective on Genetically Modified Crops." In *Environmental Virtue Ethics*, edited by Philip Cafaro and Ronald Sandler, 215–32. Lanham, Md.: Rowman and Littlefield.

Schulze, R. 2005. *Carolina Gold Rice: The Ebb and Flow of History of a Lowcountry Cash Crop*. Charleston, S.C.: History Press.

Shiva, Vandana. 2012. "Women's Indigenous Knowledge and Biodiversity." In *Environmental Ethics: Readings in Theory and Application*, edited by Louis P. Pojman and Paul Pojman, 383–88. 6th ed. Boston: Cengage Learning.

Shotwell, Alexis. 2016. *Against Purity: Living Ethically in Compromised Times*. Minneapolis: University of Minnesota Press.

Slow Food Foundation for Biodiversity. 2019. "Carolina Gold Rice: Ark of Taste." October 8. https://www.fondazioneslowfood.com/en/ark-of-taste-slow-food/carolina-gold-rice/.

Stone, Glenn D., and Dominic Glover. 2016. "Disembedding Grain: Golden Rice, the Green Revolution, and Heirloom Seeds in the Philippines." *Agriculture and Human Values* 34 (1): 87–102.

Tang, Guangwen, Yuming Hu, Shi-an Yin, Yin Wang, Gerard E. Dallal, Michael A. Grusak, and Robert M. Russell. 2012. "β-Carotene in Golden Rice Is as Good as β-carotene in Oil at Providing Vitamin A to Children." *American Journal of Clinical Nutrition* 96 (3): 658–64.

Tibbetts, John. 2006. "African Roots, Carolina Gold." *Coastal Heritage* 21 (1). https://www.scseagrant.org/african-roots-carolina-gold/.

Tuana, Nancy. 2008. "Viscous Porosity: Witnessing Katrina." In *Material Feminisms*, edited by Stacy Alaimo and Susan Hekman, 120–56. Bloomington: Indiana University Press.

Warren, Karen. 1990. "The Power and the Promise of Ecological Feminism." *Environmental Ethics* 12 (2): 125–46.

Whatmore, Sarah. 2002. *Hybrid Geographies: Nature Culture Spaces*. London: Sage.

Whitehead, Alfred N. 1964. *The Concept of Nature: Tarner Lectures Delivered in Trinity College, November 1919*. Cambridge: Cambridge University Press.

Epilogue

ONION KUCHEN
Make a flaky dough or puff pastry. Lay it very thinly on a greased baking sheet. Now stew about 10 [sliced] onions in 15 decagrams fat without burning them. Add 1 spoon pounded caraway seeds, a pinch of salt, 2–3 pieces sugar [tablets]. Add 3 spoons flour and set aside. When cool, add 3 yolks, 1 glass thick sour cream, and [egg whites, stiffly beaten] snow. Spread this on the dough layer and bake it in a medium hot oven and cut into small squares.

—FROM *IN MEMORY'S KITCHEN: A LEGACY FROM THE WOMEN OF TEREZIN*

LOVE AND LOSS
Go, go if you want to go
It's a shame, alone you will be returning
It's a shame, alone you will be returning
How I have loved that dove
How I have loved that dove
It's a shame that I am not able to forget
It's a shame that I am not able to forget
Continue, continue on your journey
Go on, go on, continue on your journey
It's a shame, I am not able to forget you
It's a shame, I am not able to forget you
Remembering the care and love
It does not matter that he/she leaves,
alone he/she will return

—QUECHUA SONG ON LOVE AND LOSS FROM POTATO PARK IN CUSCO, PERU

Embodiment. Sensuousness. Materiality. Affect. Intermingling. Cocompletion. Collaborators and interlocutors. These were notions I did not grow up with academically. I, instead, grew up with Cognition. Taxonomic trees. Decision-making models. Natural resource management. Sustainable development. Informants and participants. Reflecting on this volume, *Embodying Biodiversity: Sensory Conservation as Refuge and Sovereignty*, I ask myself, how did I—indeed, as we say around where I am, how did "thangs"—get from there to here?

There were changing currents in the anthropological literature and research trajectories. Heady and far-reaching, these turns blurred my focus from cognition and categorization to landscape and place. There were field experiences that shaped, and continue to shape, my consciousness; an inescapable feeling of deep and widening connection, an abiding appreciation for the gentle audacity at the margins vis-à-vis the harsh light and self-serving illusions of modernity, an ongoing negotiation with agency and indeterminacy. There were life experiences that brought to the fore latent tendencies of resistance and resilience, the need to thrive in place, or to root *out-of-place*, to consecrate my altars (with a lowercase *a*) wherever I landed.

But above all, the inspiration to move on this journey came from my students and the larger community of younger colleagues I had the privilege of knowing. They filled their bibliographies with rhizomatic citations that penetrated geographic and theoretical outposts. They came back from the field where "the sun shone all the time," "the women cooked soup all day," and "people, young and old, kept blessing you!"; where breastfeeding women navigated between local knowledge of plants and animals that made their milk flow and government messages on the clinical, and therefore proper, way of maternal and child care; where the rustic grapes chose to grow on volcanic soil; where locals and migrants had to craft creative relations with plants and microbes to keep what is vital in the face of temporal and spatial displacements. Individually and collectively, we vacillated between "People and Their Plants" and "Plants and Their People." Through undergraduate and graduate seminars and professional workshops and conferences, as well as visits and productive email exchanges, I met interesting individuals working on the verges of disciplines, exploring methodological and substantive borderlands.

A good representation of these younger colleagues' work is found between the covers of this edited volume. Sensory conservation is conservation with a small *c*, on the ground and in small spaces. The first part is on seeking, or *making*, refuge. Against the violence of forgetting, people in various locations, and occasionally in different locations at the same time, have stuck to something concrete and usually portable—in their farms and gardens, kitchens, places of worship, and in refuge camps—to get them through extreme hardship and dehumanizing placelessness. As these concrete anchors become more tenuous or inaccessible, they still hold on to them in their minds and hearts, much like women in the concentration camp of Terezin sustained themselves through hunger and diminishment with memories of the pots and pans in the kitchens they left behind, the smell of rich chocolate cake, the complex pungency of the onion kuchen.

That these movable anchors to an affective homeland consist mostly of plants is no surprise given the self-selection of this community of scholars, but it also comes from how well plants are suited to make this yearning tangible. Plants are cognitively salient; their sharp leaves scratch and cut into you as you traverse the rice paddies, as the landless workers in Kabaritan, a rice-farming community in the Philippines where I did my dissertation fieldwork, pointed out. They are packed and sensuous, their appearance, smell, taste, and texture both grounding and transporting, as contributors to this volume have so powerfully evoked. They summarize what is good, as small-scale farmers in Bukidnon, Philippines, refer to the beautiful (*guapo kaayo*, or very handsome) sweet potatoes, and congeal what is precious, as Quechua farmers in the Peruvian Andes cradle their potato *wawas* (infants). They are also gifts of commensality, conveying caring and concern in the sharing of food. All in all, it is a subterranean love affair that sustains our humanity.

In the second part of the volume, the authors explore inscribed and embodied worlds and the tension that comes with *sovereigning the sensuous.* They tackle the question of reconciling the "underground" plants with the reckoning of heft that power demands. More importantly, they shed light on how this has been done, and is being done, in fields, markets, and laboratories that are not necessarily in out-of-the-way places or on the edges of things. The fact is, people are countering loss by weaving intimate/animate landscapes, whether concrete or for the time being imaginary, across time and space. Like the Quechua farmers who swaddle and scold their potatoes, they sing back what is loved for they cannot help remembering despite the "shame." Sometimes celebratory, often nonchalant, these spaces of contest—or noncontest—are minute, diverse, and articulate *sovereigntiae.*

The conclusion of this collection, and its significant and I hope enduring contribution, is that the relationship between persistence and biodiversity, even more than the conservation of biodiversity per se, is critical to ensuring survival and understanding much of what goes on around us. Against dislocation, a problem that haunts not only migrants and refugees but also those displaced by time, humans strive to be "rootfull," to *return* to what gives them a sense of comfort, pleasure, and belonging, and conservation comes as a hefty bonus. Loss, like beauty, is in the eye of the beholder; we garden in a landscape of memory. As El Salvadoran milpa farmers, Indian diasporic devotees, young Karen refugees, Ugandan banana gardeners, Mapuche homegardeners, Latinx farmers, Ethiopian ensete growers, and Carolina Rice advocates remind us, affective ties bind, and it is owing to this that biodiversity flourishes. They have demonstrated that sovereignty is permission we give ourselves to care for countless cobeings no less than for oneself.

I am deeply honored by this collection and humbly thank the volume contributors, Michael Anastario, Shannon A. Novak, Terese V. Gagnon, Yasuaki Sato, José Tomás Ibarra, Julián Caviedes, Antonia Barreau, Emily Ramsey, Valentina Peveri, and Justin Simpson, and especially its editor. Terese has foraged, tethered, simmered, and served these offerings as she traversed her own landscape in kinship with the Karen and their traveling botanical companions. The chapters are stories rich in ethnographic insights and theoretical implications. Forgoing commitment to any one scholarly paradigm, these plural stories are woven with candidness and empathy, a cleaving to how humans and their kins experience and embody each other as they yearn-and-tend to what makes them whole, in other words, to what is precious in precarity. Coercion and dislocation as well as trauma and forgetting are very much a part of these stories, but so are refuges made and sovereignties reclaimed through the diversity of plants we carry, cultivate, cook, and coconsume. As we transport them, these portals of affection transport us. As we grow them, they grow us.

—Virginia D. Nazarea

THE FLAME TREE

My neighbor has decided to poison the flame tree.
He is right, of course.
The tree is over 20 years old, huge, spreading,
and the termites have worn jagged roads clear to its top.
It's clearly a danger
tilting toward our house—
some fickle wind
my neighbor says could blow it over.

Every fañomnåkan, it sends out its bursts of orange blossoms;
it blooms and blooms and blooms relentlessly,
the flares it sends shooting out into space
more stunning than fireworks
through the window

where my mother
riveted to a bed, doomed by her body to a colorless spot,
gazes out, her head on a pillow—
might have seemed like forever to her who used to climb green
    mountain sides—
and watches that tree full of sparrows
chittering
chattering
flitting here and there
and the outlandish blazing petals
steadfastly singing against the blue sky.

My neighbor, true to his word,
injected a poisonous brew bought at Home Depot into the trunk of
    the tree,
the toxic river
traveling up up up following the termite trails to the heart
of the fire.

He is right, of course.

The tree came back the following year,
its clusters unflinchingly parading their bursts of rebellious orange.
But the poison had done its work—
see, where there was a canopy of flames
there are now just a handful here and there,
one spray in particular desperately
reaching out
like a fist full of beauty
to the window

where she
used to watch for its return.

—Evelyn Flores

CONTRIBUTORS

Mike Anastario is an assistant professor of health sciences at Northern Arizona University. A former Fulbright Scholar to El Salvador and Harvard-JPB Environmental Health Fellow, he is the author of *Parcels: Memories of Salvadoran Migration* and co-author of *Kneeling Before Corn: Recuperating More-than-Human Intimacies on the Salvadoran Milpa*. Anastario's research interests concern rural health disparities and the science of harm reduction.

Antonia Barreau is an ethnoecologist working at Ecosystem-Complexity-Society (ECOS) Co-Laboratory and Center for Local Development at the Pontificia Universidad Católica de Chile. Her research focuses on community–forest relations, food sovereignty, and homegardens. Antonia is the founder of Del Monte a la Cocina, an independent project working with forest wild edibles and local cuisine. She is also a member of the Chilean Society of Socioecology and Ethnoecology.

Julián Caviedes is a PhD student at the Institut de Ciència i Tecnologia Ambientals of the Universitat Autònoma de Barcelona. He is a researcher at ECOS Co-Laboratory and Center for Local Development, Villarrica Campus, Pontificia Universidad Católica de Chile. Julián is also a member of the Chilean Society of Socioecology and Ethnoecology. His recent research interests are focused on agroecology, food sovereignty, and the resilience of campesino and Indigenous social-ecological systems.

Terese V. Gagnon is a political and environmental anthropologist. She is a postdoctoral fellow at the University of North Carolina, Chapel Hill, with the Bringing Southeast Asia Home Initiative. She holds a PhD in anthropology from Syracuse University and is currently writing a book about Karen food, seed, and political sovereignty across landscapes of home and exile. She is co-editor, with Virginia Nazarea, of *Movable Gardens: Itineraries and Sanctuaries of Memory*. Terese incorporates creative forms, including ethnographic poetry and visual anthropology, in her scholarly work.

José Tomás Ibarra is a lecturer and researcher on human ecology and social-ecological systems at ECOS Co-Laboratory, Center for Local Development, and Center for Intercultural and Indigenous Research, Villarrica Campus and Faculty of Agriculture and Forestry Sciences, Pontificia Universidad Católica de Chile; Center of Applied Ecology and Sustainability; and Cape Horn International Center for Global Change Studies and Biocultural Conservation. Tomás is also a member of the Chilean Society of Socioecology and Ethnoecology. His research interests span from human ecology through agroecology to transdisciplinarity, with a focus on social-ecological systems such as forests, agroforests, mountains, and small-scale agriculture.

Shannon A. Novak is a professor of anthropology at Syracuse University. Her research seeks to understand human bodies as living organisms, material entities, and cultural symbols. Broadly trained in bioarchaeology and social anthropology, Novak's studies engage with ethnoreligious movements across the long nineteenth century and their continued resonance in more-than-human landscapes and embodied practices. Her current project follows a healing cult carried by indentured laborers from India to the cane fields of British Guiana in South America and, in recent decades, by Indo-Guyanese immigrants to North American cities. Since 2016, she has been conducting ethnographic research with one such community in Brampton, Ontario.

Valentina Peveri is a social anthropologist with interests in political ecology. Since 2004 she has maintained an energetic connection with her fieldwork in southern Ethiopia. She held a Fulbright and visiting scholar appointment at Boston University and was awarded a Hunt postdoctoral fellowship by the Wenner-Gren Foundation for the creation of a book entitled *The Edible Gardens of Ethiopia: An Ethnographic Journey into Beauty and Hunger* (2020). She has published and edited several books, among them *L'albero delle donne: Etnografia nelle piantagioni e cucine d'Etiopia* (2012), and also many chapters on her food, environment, and gender research. Her strong ethnographic background has been complemented over time by appointments as international consultant with a special focus on the role gender and other axes of power play in natural resource and environmental policies. She currently serves as an adjunct professor for the MA in food studies at the American University of Rome.

Emily Ramsey is a PhD candidate at the University of Georgia broadly interested in small farming and organic agriculture in the United States, as well as in the way knowledge systems, memory, and agricultural heritage inform how individuals preserve farming practices, maintain traditional varieties, and pass knowledge on to generations. She is particularly interested in the rapidly growing yet still largely unheralded role that Latinx immigrants are playing in U.S. small farming as producers. Emily is interested in how memories of transnational migrants carry over into the agricultural practices and knowledges they may wish to preserve.

Yasuaki Sato is associate professor at Nagasaki University, where he teaches the relationship of culture and biodiversity. His ethnobotanical research examines the livelihood system based on bananas, especially in East Africa and Papua New Guinea. He is a co-author of *Cooking Banana in Africa* and the author of *Life-World of Banana Cultivators in Uganda: An Ethnoscience Approach* (in Japanese). His current project focuses on dietary education in consideration of local wisdom in rural Uganda through a participatory method.

Justin Simpson is a lecturer in philosophy at the University of Texas, Rio Grande Valley. He received his PhD in philosophy from the University of Georgia in 2022. His research cross-pollinates posthumanisms and new materialisms with environmental ethics and social epistemology. He has previously published on the significance of contingency, distance, and detours in the philosophical anthropology of Hans Blumenberg; the positive virtues of an active listener in eventful conversations; and advancing Robin Wall Kimmerer's joyful environmental ethic of reciprocity through the virtue of open-ended curiosity. He is currently co-authoring a book that argues for playful and loving attention as an environmental virtue in a world of surprising nonhumans.

INDEX